Genome Refactoring

Synthesis Lectures on Synthetic Biology

Editor

Martyn Amos, Manchester Metropolitan University

Genome Refactoring
Natalie Kuldell and Neal Lerner
Massachusetts Institute of Technology

Concepts and Engineering in Synthetic Biosensing/-reporting Microorganisms
Jan Roelof van der Meer
University of Lausanne

Genome Refactoring
Natalie Kuldell and Neal Lerner

ISBN: 978-3-031-01441-3 paperback

ISBN: 978-3-031-02569-3 ebook

DOI: 10.1007/978-3-031-02569-3

A Publication in the Springer series

SYNTHESIS LECTURES ON SYNTHETIC BIOLOGY # 1

Lecture #1

Series Editor: Martyn Amos, Manchester Metropolitan University

Series ISSN pending

Genome Refactoring

Natalie Kuldell and Neal Lerner
Massachusetts Institute of Technology

SYNTHESIS LECTURES ON SYNTHETIC BIOLOGY # 1

ABSTRACT

The science of biology celebrates the discovery and understanding of biological systems that already exist in nature. In parallel, the engineering of biology must learn how to make use of our understanding of the natural world to design and build new useful biological systems. "Synthetic biology" represents one example of recent work to engineer biological systems. This emerging field aims to replace the *ad hoc* process of assembling biological systems by primarily developing tools to assemble reliable-but-complex living organisms from standard components that can later be reused in new combination. The focus of this book is "genome refactoring," one of several approaches to manage the complexity of a biological system in which the goal is to redesign the genetic elements that encode a living form—preserving the function of that form but encoding it with a genome far easier to study and extend. This book presents genome refactoring in two ways: as an important aspect of the emerging field of synthetic biology and as a powerful teaching tool to train would-be professionals in the subject. Chapters focus on the overarching goals of synthetic biology and their alignment with the motivations and achievements in genome engineering; the engineering frameworks of refactoring, including genome synthesis, standardization of biological parts, and abstraction; a detailed description of the bacteriophages that have been refactored up to this point; and the methods of refactoring and contexts for that work drawn from the bacteriophage M13. Overall, these examples offer readers the potential for synthetic biology and the areas in need of further research. If successful, synthetic biology and genome refactoring could address any number of persistent societal needs, including sustainable energy, affordable and effective medicine, and green manufacturing practices.

KEYWORDS

synthetic biology, genome refactoring, abstraction, standardization, modularity, DNA synthesis, bacteriophage, T7, M13, genetic parts

Preface

The science of biology celebrates the discovery and understanding of biological systems that already exist in nature. In parallel, the engineering of biology must learn how to make use of our understanding of the natural world to design and build new useful biological systems. *Synthetic biology* represents one example of recent work to engineer biological systems. This emerging field aims to replace the ad hoc process of assembling biological systems, primarily by developing tools to reliably assemble complex living organisms from standard components that can later be reused in new combination. Early examples of synthetic biological systems include engineered bacterial surface proteins that recognize materials such as TNT (Looger, Dwyer, Smith, & Hellinga, 2003), engineered metabolic pathways that produce drug precursors such as artemisinin (Keasling, 2008), and primitive memory and logic elements such as latches and ring oscillators that form synthetic genetic circuits (Elowitz & Leibler, 2000; Gardner, Cantor, & Collins, 2000). If successful, synthetic biology could address any number of persistent societal needs, including sustainable energy, affordable and effective medicine, and green manufacturing practices.

One obstacle to the reliable and scalable engineering of living systems is that biology has been optimized by evolution, resulting in genetically programmed "machines" that are robust and elegant but also difficult to understand and even more difficult to manipulate. *Genome refactoring* is one of several approaches to manage the complexity of biological system, and this approach will be the focus of this lecture. The goal of genome refactoring is to redesign the genetic elements that encode a living form—preserving the function of that form but encoding it with a genome far easier to study and extend.

This book aims to present genome refactoring in two ways: as an important aspect of the emerging field of synthetic biology and as a powerful teaching tool to train would-be professionals in the subject. In Chapter 1, some overarching goals of synthetic biology are aligned with the motivations and achievements in genome engineering and refactoring. Engineering frameworks for managing the complexity of biology are described, including a description of genome synthesis, standardization of biological parts, and abstraction. In Chapter 2, the bacteriophages that have been refactored are detailed. Chapter 3 focuses on the methods and outcomes of two reengineering efforts, and Chapter 4 describes the usefulness of synthetic biology for teaching professional practices to biological engineering students. In Chapter 5, the struggles and successes of refactoring are revisited in terms of the scientific understanding they engender as well as the engineering end points they facilitate.

REFERENCES

Elowitz, M. B., & Leibler, S. (2000). A synthetic oscillatory network of transcriptional regulators. *Nature, 403*(6767), pp. 335–338.

Gardner, T. S., Cantor, C. R., & Collins, J. J. (2000). Construction of a genetic toggle switch in *Escherichia coli. Nature, 403*(6767), pp. 339–342.

Keasling, J. D. (2008). Synthetic biology for synthetic chemistry. *ACS Chemical Biology, 3*(1), pp. 64–76.

Looger, L. L., Dwyer, M. A., Smith, J. J., & Hellinga, H. W. (2003). Computational design of receptor and sensor proteins with novel functions. *Nature, 423*(6936), pp. 185–190.

Acknowledgments

This lecture is a direct result of our productive and enjoyable time teaching with Drew Endy. He is an extraordinary colleague. We also gratefully acknowledge the students in our undergraduate subject, Laboratory Fundamentals of Biological Engineering. Their enthusiasm, creativity, and talent are exceptional. Finally, many thanks to Martyn Amos and Diane Cerra for their generous support of this project.

Contents

CHAPTER 1

Tools for Genome Engineering and Synthetic Biology

1.1   INTRODUCTION

Three foundational techniques—all now at least 30 years old—enable genetic engineering. These techniques are recombinant DNA methodologies, DNA sequencing, and PCR (Figure 1.1). This suite of tools allows working scientists to move, read, and copy DNA. They make commonplace a researcher's ability to molecularly manipulate genetic material and to rationally shuttle genes from one organism to another. Yet nearly any working scientist who relies on these techniques can attest to time lost when DNA will not recombine as desired for reasons that usually remain obscure. Alternatively, weeks can be spent planning and then executing the needed manipulations, only to generate what turns out to be unstable or nonfunctional genetic material. Thus, although the practice of using restriction endonucleases, ligases, plasmid vectors, PCR, and sequencing might collectively be termed *genetic engineering*, it currently fails to fulfill the profile of mature engineering efforts, which feature reliable and scalable foundations.

Some effort to advance the work of synthetic biology relies on the application of foundational engineering principles to the building of living systems. Drawing lessons from more mature engineering fields such as civil, electrical, and mechanical engineering, synthetic biologists are learning to define biological standards and manage system complexity through abstraction. Standardization and abstraction are integral features of mature engineering disciplines. Indeed, they are nearly taken for granted. Yet these foundational engineering practices have not been applied to living systems in any concerted fashion. We begin this lecture by exploring their usefulness for genome engineering in particular and to synthetic biology more broadly.

Natural biological systems are evolved, and evolution does not produce designs that are easy for humans to understand. This point is particularly well illustrated by the scientific study of bacteriophage. Molecular-level understanding of these simple living systems has come after 50 years or more of genetic, biochemical, and sequence analysis, requiring extensive and thoughtful work across many labs (Calendar, 2006). Yet computer models based on the collected data inadequately

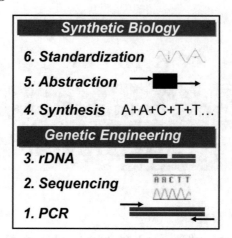

FIGURE 1.1: Tools for genetic engineering and synthetic biology.

predict bacteriophage behavior, suggesting that a complete understanding of these "simple" systems remains elusive (Endy, You, Yin, & Molineux, 2000).

Synthetic biology offers an alternative approach to the study of these phages. Refactored genomes, being human rather than natural designs, might be the next technical advance to better understand bacteriophage as natural biological systems. T7 and M13 are the two bacteriophages to which refactoring has been deployed (Chan, Kosuri, & Endy, 2005; Kuldell, 2007). The work with these phages illustrates how genome-scale engineering can serve and extend scientific endeavors by (1) providing surrogate systems to assess the descriptions of the natural one and (2) offering "user-friendly" templates for tasking the natural systems to human-defined applications.

Early and limited efforts at genome refactoring provide both insight and inspiration. Scientifically speaking, refactoring efforts highlight the fact that we still have an incomplete understanding of these phages despite a generation of study. If scientists further engage in genome refactoring as a tool for exploring the limits and errors in our understanding, then we can hope to better describe the rules and mechanisms that govern the natural instances (T. F. Knight, 2005). As an engineering practice, genome refactoring can help build a library of genetic "parts," snippets of DNA that perform defined functions (Endy, 2005). Such a collection of parts will populate an engineer's toolkit and someday enable the design and synthesis of novel genomes, assuming we gain enough savvy to know how parts and combinations of parts will behave in a living cell. The clearest and most immediate lesson from refactoring efforts, however, is this: Biology is a task-able technology. The better we become at rational design and deliberate construction of genomes, the better we will be at deciphering the underlying mechanisms that guide natural and artificial constructs and the faster we will be able to build useful, robust, novel biological systems.

However, genome refactoring is only one of the tools being deployed by biological engineers. Before considering the particulars of the T7 and M13 examples in Chapter 2, some general frameworks for synthetic biology and the engineering of living systems need to be established.

The ability to read, write, and copy DNA has elucidated genetic details of living systems and has provided the means of constructing relatively simple genetic programs. Synthetic biology is wholly dependent on existing techniques of molecular biology but applies other engineering and technological foundations to further develop biology. To make living systems a more taskable technology, synthetic biologists deploy (1) gene- and genome-scale synthesis of DNA, (2) abstraction, and (3) standards to make intelligent use of sequence information. We now describe each of these tools/frameworks in more detail.

1.2   DNA SYNTHESIS

The first of these tools, solid-phase gene and genome synthesis, is rapidly enabling a new and complementary approach to the more common tools of assembling genetic information from existing DNA templates. The de novo synthesis of DNA through phosphoramadite chemistry was described by Beaucage and Caruthers (1981) and improved with tetrazole catalysis in 1983 (Caruthers et al., 1983). These publications detailed the methods for sequentially adding a desired nucleotide unit to an immobilized polynucleotide chain (Figure 1.2). The chemistry was initially applied to synthesize short oligonucleotides, on the order of 20–60 bases, and these molecules have since been pervasively used for molecular techniques such as sequencing and PCR.

The cost for solid-phase synthesis of short oligonucleotides has been decreasing by a factor of 2 every 18 months (Figure 1.3). This trend has held since 1990 (Carlson, 2008). Decreasing synthesis costs are coupled with increasingly efficient assembly reactions. Success rates of approximately 99.5% per nucleotide addition are now commonplace (Integrated DNA Technologies, 2005), and assembly of gene-length DNA is both possible and increasingly affordable. Indeed, the cost of gene-length synthesis is currently decreasing even faster than the cost for oligonucleotide synthesis (0.3 orders of magnitude per year for genes versus 0.2 orders of magnitude per year for the already-inexpensive oligonucleotides).

Continued progress in gene- and genome-length synthesis can be powerfully coupled with the information in sequence databases (Bio FAB Group et al., 2006). Physical DNA can be compiled from digital Gs, As, Ts, and Cs, circumventing the need for even a single copy of a genome or genomic element. In addition, the combination of sequence data and synthesis technology allows for the writing of previously unwritten genetic code, and one can imagine a day when synthesis of DNA will be sufficiently fast and inexpensive to enable any combination of genes to be specified, constructed, and tested. The workings of a genome could then be probed using high-throughput construction and testing of variants, a research approach that is appealing but currently too costly and time-intensive.

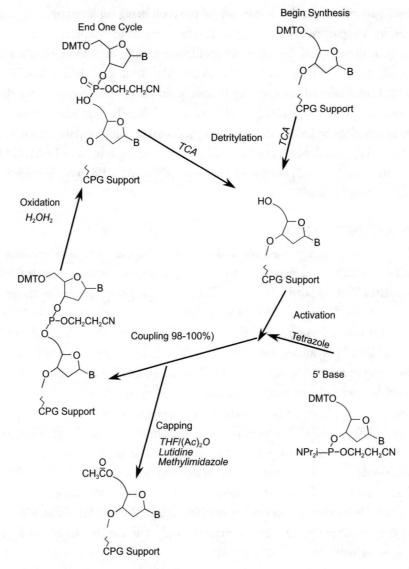

FIGURE 1.2: DNA synthesis chemistry.

The future application of synthetic genomes as a research tool depends on improvements in DNA synthesis technology as well as on improved rules for writing novel genomes. Currently, we have inadequate understanding of the architecture and the boundaries of genomic parts to confidently write all new genomes. Even the genomes of simple entities like bacteriophage are complex in their architecture, and despite years of study, we do not fully appreciate the functions of numerous sequence elements. Our limited fluency in genomic syntax impinges on our ability to wisely design

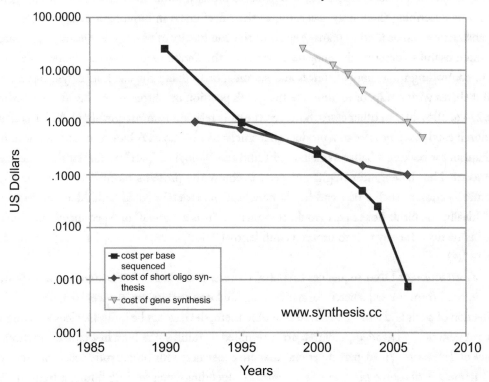

FIGURE 1.3: Cost of synthesis.

novel genomes. However, as detailed below, gains in fluency may be realized through engineering principles now being applied to biology, namely, abstraction and standards.

1.3   ABSTRACTION AS A TOOL FOR MANAGING AND UNDERSTANDING BIOLOGICAL COMPLEXITY

A second tool of synthetic biology that may facilitate the engineering of living systems is abstraction. Abstraction is a common engineering practice to manage complexity and lies between the fully reductionist approach to biology, which hopes to describe complex living systems through their underlying chemistry and physics, and the exclusively holistic approach, in which every component is connected. Neither the reductionist nor the holistic extreme has yet to fully support the rational design of desired biotechnologies (Van Regenmortel, 2004).

Abstraction removes any reference to a specific instance of a function and represents only the underlying concept or idea that unifies several examples. In computer science and software design, abstraction allows designers to factor out details so they may control them, dealing with fewer concepts at a time. Abstraction, as applied to biology, limits the number of relevant genome features by describing them only according to their usefulness in building a living system. In this way, abstraction can be used to manage some of the complexity of biology. For example, researchers can find it useful to describe the enzyme encoded by the *Saccharomyces cerevisiae* gene ATF1 for its role in controlling the export of sterols and steroids, or to detail the alcohol acetyltransferase domain it shares with Atf2p, or to annotate the gene's position on chromosome XV from coordinates 1046224 to 1047801. In other cases, however, the only relevant information about ATF1 is that the reaction it catalyzes gives rise to a product that smells like bananas. A bioengineer, hoping to find a mechanism for making a banana smell, might find this biological "part" useful but the other details extraneous. That bioengineer might not need to know the protein's motif details or the original genomic location of such a part and would benefit if these details could be hidden within a "black box." Ideally, the black-boxed part would come with a "user's manual" or "spec sheet" that included basic instructions for any user to proceed confidently (Arkin, 2008; Canton, Labno, & Endy, 2008) (Figure 1.4).

Abstraction of DNA sequences to "parts" represents the base of an abstraction hierarchy, one step removed from the sequence information itself and one step below "devices" (Figure 1.5). With a collection of parts and a mechanism to assemble them, devices can be built. Devices are composed of $N > 1$ genetically encoded parts that are assembled to fulfill some human-defined function. Examples of devices are "quad-part inverters" that are constructed for prokaryotic cells from four basic parts, namely, a ribosome binding site, a repressor-encoding open reading frame, a transcriptional terminator, and a promoter that is repressed by the encoded repressor. These devices are intended to convert high-input signals to low, and low to high, while hiding the details of precisely how they are accomplishing this task. A second example of a device is a "reporter construct," also assembled from four basic parts and intended to operate in bacteria. Reporter devices consist of a promoter part, a ribosome binding site, an open reading frame that generates a measurable product (Green Florescent Protein, or GFP, for example), and a transcriptional terminator. Depending on the needs of the bioengineer, prokaryotic devices can be assembled according to these recipes but from different component parts. Eukaryotic devices with similar functions can likewise be genetically encoded by assembling basic parts, although largely requiring different basic parts and different composition recipes. Independent of intended cellular host, the abstraction of the parts according to their functions, along with the natural modularity of biological functions, enables a library of devices to be constructed. From an annotated collection of parts and device, other bioengineers can rationally select a device based on its past performance and appropriate it to fit a new application.

The upper layer of the abstraction hierarchy is the "systems" layer. Ideally, a collection of characterized devices could be mixed and matched to generate complex cellular functions. In a perfect

BBa_F2620

3OC$_6$HSL → PoPS Receiver

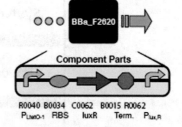

Component Parts

R0040 B0034 C0062 B0015 R0062
P$_{LtetO-1}$ RBS luxR Term. P$_{luxR}$

Mechanism & Function

A transcription factor (LuxR) that is active in the presence of a cell-cell signaling molecule (3OC$_6$HSL) is controlled by a regulated operator (P$_{LtetO-1}$). Device input is 3OC$_6$HSL. Device output is PoPS from a LuxR-regulated operator. If used in a cell containing TetR then a second input such as aTc can be used to produce a Boolean AND function.

Static Performance*

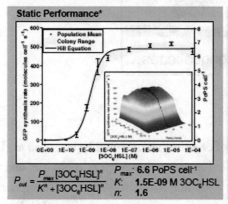

$$P_{out} = \frac{P_{max}[3OC_6HSL]^n}{K^n + [3OC_6HSL]^n}$$

P_{max}: 6.6 PoPS cell^{-1}
K: 1.5E-09 M 3OC$_6$HSL
n: 1.6

Dynamic Performance*

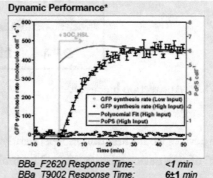

BBa_F2620 Response Time: <1 min
BBa_T9002 Response Time: 6±1 min
Inputs: 0 M (Low), 1E-07 M (High) 3OC$_6$HSL

Input Compatibility*

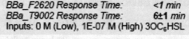

Part Compatibility (qualitative)

Chassis: MC4100, MG1655, and DH5α
Plasmids: pSB3K3 and pSB1A2
Devices: E0240, E0430 and E0434

Transcriptional Output Demand (low/high input)

Nucleotides: 0 / 6xNt nucleotides cell^{-1} s^{-1}
Polymerases: 0 / 1.5E-1xNt RNAP cell^{-1}
(Nt = downstream transcript length)

Reliability**

Genetic: >92/>56 culture doublings
Performance: >92/>56 culture doublings
(low/high input during propagation)

Conditions (abridged)

Output: PoPS measured via BBa_E0240
Culture: Supplemented M9, 37°C
Plasmid: pSB3K3
Chassis: MG1655
*Equipment: PE Victor3 multi-well fluorimeter
**Equipment: BD FACScan cytometer

Signaling Devices

http://parts.mit.edu/registry/index.php/Part:BBa_F2620

Authors: Barry Canton
Ania Labno
Updated: March 2008

Registry of Standard Biological Parts
making life better, one part at a time

License: Public

FIGURE 1.4: Parts specification sheet.

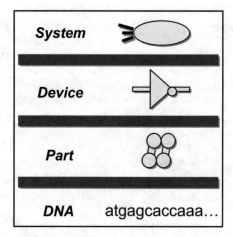

FIGURE 1.5: Abstraction hierarchy.

world, the devices would function independently of one another or would be insulated to eliminate "cross talk" of one device with another. The limited number of regulatory molecules and tunable promoters that are currently available makes such insulation mandatory, even as bioengineers build relatively simple genetic programs. It is possible that as more genomes are sequenced and more parts are defined, there will be a sufficient number of orthogonal components to design complex systems that behave predictably.

The abstraction hierarchy encourages the enumeration of many biological parts, as well as the reliable composition of these parts into biological devices and then systems. It also requires that the protocols for interfacing the abstraction layers be defined and sensible. There are well-publicized disasters in other engineering disciplines that can be traced to poor protocols for communication between relevant but decoupled facets of a project. The collapse of the walkways at the Kansas City Hyatt Regency in 1981 is a sobering example of an engineering failure arising from poor interfacing of the designers and the builders (Petroski, 1985). By calling for a beam that was longer than could possibly be transported on standard trucking equipment, the designers considered only the goals within their layer of the system-building hierarchy. When it came time to implement the system, the builders were forced to substitute two shorter beams that could be transported and improvised a patchwork mechanism for holding the two shorter beams together to support the suspended walk-ways. Investigation of the walkway collapse showed that the failure of the patch was predictable but that, in the absence of defined communication protocols between the designers and the builders, neither of the individuals could be found at fault. It is notable, in this regard, that the designers of the first refactored T7 genome, unlike the designers of the Kansas City Hyatt, considered the "buildability" of the (genome) parts they specified (Chan et al., 2005; Kosuri, 2007).

1.4 STANDARDS

Complete genome sequences are readily available for many interesting living forms, profoundly changing genetic analysis. Before the availability of complete sequencing data, genetic studies identified mutant phenotypes and then used recombination frequency to map and characterize the relative genetic elements. Later, with genomic sequence data available, information based on sequence patterns revealed additional genome features. The genes and regulatory elements identified on the T7 genome are a case in point (Dunn & Studier, 1983; Studier, 1969). Genomic sequencing data allow putative genetic elements to be identified based on their resemblance to other genetic features in other organisms. Given any newly sequenced genome, a pattern recognition model can identify the genetic "parts" within.

However, even an extensive collection of genetic parts will not transition genetic engineering away from ad hoc projects because they will still require clever strategies for cutting and pasting the DNA to build genetic devices. Rather, the parts need to be refined according to a series of assembly and characterization rules. It is standard parts that populate the base of the abstraction hierarchy described in the preceding sections, as these standard parts support the upper layers of the hierarchy by enabling fast, reliable compilation of genetically encoded devices and systems. Thus, although a list of basic biological parts can arise naturally from gains in genetics and genomics research, a *standard* parts list requires a more self-conscious effort aimed at refining the parts themselves. Parts can be refined to meet a standard that facilitates physical or functional composition of the parts into working devices.

Standards are a necessary aspect of engineering. In the absence of established standards, machines cannot talk to each other and hardware is difficult to repair. Two illustrative examples are frequently cited (Brockman & Weinberger, 2008). First, consider the transition in American machine shops that occurred in the 1860s. Before that time, it was commonplace for machine shops to build materials with parts that met only personal specifications. This resulted in machines that could be repaired only at the original point of manufacturing because the screwthreads that held bolts to nuts were not compatible with those parts from other shops. William Sellars from the Franklin Institute in Philadelphia proposed a standard angle and shape for screwthreads, and the standard was accepted. Suddenly, parts were interoperable and far easier to deploy and distribute. Success required that the screwthread standard be accepted and that machine shops retool their equipment to manufacture parts accordingly. After this initial investment to meet an accepted standard, all machinists benefited from the simplified fixing and building cycles.

The second often cited example to illustrate the usefulness of standards begins before the American Civil War, when railway companies purposefully used different gauges on their train tracks to inhibit shippers from switching lines. This lack of uniformity in the spacing between the train tracks became a nightmare during the Civil War. Destruction of a short passage of track meant equipment and troops were unmovable, and so when railroads were reconstructed, they were built with standard gauges, thereby allowing working tracks to substitute for broken ones. George Stephenson rallied support for the now standard rail gauge of 4 feet 8.5 inches (1435 mm), and his contribution to conformity has been recognized with a unit in his name, the "Stephenson gauge."

Taken together, the screwthreads and the railway gauge highlight motivations and consequences of standardization. Genome refactoring efforts can be viewed as a similar step to conformity within the investigative community, greatly diminishing the effort needed to share genetic elements and discoveries. A standard architecture for genomes might yield a DNA template that does not require great artistry or savvy to work with or multiple control experiments to understand. Just as a degree in mechanical engineering is not needed to lock a nut on a bolt, refactored genomes should be composed of standard parts that can be used intuitively or after a brief familiarization.

By necessity, standardization requires some features and properties of DNA to be limited, and the decision of which aspects to limit will depend on what application the standards are intended to support. A common, although not universal, standard for biological parts is the "BioBrick" standard (T. Knight, 2003). Sequence features of "BioBrick-ed" parts are limited to enable their reliable physical assembly. Any two BioBricks can be assembled using the same cloning strategy. Moreover, the resulting composite BioBrick-ed parts can be starting material for further BioBrick assemblies, allowing complex DNA constructions to be performed using a single, established strategy. For example, the part described earlier that generates the banana scent has been refined to include prefix and suffix sequences for molecular cloning and given the part number BBa_J45014 (Registry of Standard Biological Parts, 2009). To assemble J45014 into a device that expresses in *Escherichia coli*, a standard promoter (e.g., BBa_I12007) and a standard RBS (e.g., BBa_J61100) can be added. Alternatively, a genetic inverter device such as BBa_Q04121 can control the banana odor part. To be clear, however, it is the application of standards to both the genetic components themselves and to their assembly that is novel and not the ability to manipulate and design DNA and control its expression. Yet as the collection of standard parts and devices grows, the barriers to building novel constructs should diminish.

1.5   GENOME REFACTORING IN THE SERVICE OF SYNTHETIC BIOLOGY

It has been predicted that DNA sequencing and synthesis technologies will be sufficiently advanced within a few years that a single person will be able to sequence or synthesize 10^{10} bases a day (Carlson, 2003). It is also been observed that information in public sequence databases like NCBI is doubling every 18 months (Stahler, Beier, Gao, & Hoheisel, 2006). However, making good use of the sequence data and synthesis technology is a challenge. Synthetic biology is currently hampered by slow and costly design cycles. Until emergent properties of designed systems can be eliminated or predicted, the engineering of biology will remain an ad hoc effort with some notable successes and some inexplicable failures.

One strategy for addressing this challenge and advance the field of synthetic biology is to refactor genomes as the natural sequences become available. Refactoring might enable natural sequences to be parsed into standardized, reliable, composable components that can predictably assemble into functional devices and systems. The parts might be tagged or somehow modified to enable their measurement and characterization. Elements of unknown function in the natural sequence might be actively removed from the new genome. Taken a step further, genomes could be "minimized" to function under a more restricted set of conditions. Chemical synthesis of DNA variants also allows genome-wide shuffling experiments that are currently labor-intensive to carry out and nearly impossible to interpret given the overlapping structure and redundant functions of many genomic elements.

What is described in the chapters that follow are two initial efforts at genome refactoring. Genome-scale engineering of T7 and M13 are instructive both for the strategies and challenges they reveal and also for the teaching tools they provide as we train the next cadre of biological engineers. As we consider the redesign of bacteriophage genomes, it is worth noting that more complex genomes encoding more complex "gadgets" are on the refactoring horizon. There is much to learn from these early efforts and much still to do.

REFERENCES

Arkin, A. (2008). Setting the standard in synthetic biology. *Nature Biotechnology, 26*(7), pp. 771–774. doi:10.1038/nbt0708-771

Beaucage, S. L., & Caruthers, M. H. (1981). Deoxynucleoside phosphoramidites—A new class of key intermediates for deoxypolynucleotide synthesis. *Tetrahedron Letters, 22*(20), pp. 1859–1862. doi:10.1016/S0040-4039(01)90461-7

Bio FAB Group, Baker, D., Church, G., Collins, J., Endy, D., Jacobson, J., et al. (2006). Engineering life: Building a fab for biology. *Scientific American, 294*(6), pp. 44–51.

Brockman, J., & Weinberger, R. (2008). *ENGINEERING BIOLOGY: A talk with Drew Endy.* Retrieved March 18, 2009, from http://www.edge.org/3rd_culture/endy08/endy08_index.html

Calendar, R. (Ed.). (2006). *The bacteriophages* (2nd ed.). Oxford: Oxford University Press. doi: 10.1086/509419

Canton, B., Labno, A., & Endy, D. (2008). Refinement and standardization of synthetic biological parts and devices. *Nature Biotechnology, 26*(7), pp. 787–793. doi:10.1038/nbt1413

Carlson, R. (2003). The pace and proliferation of biological technologies. *Biosecurity and Bioterrorism: Biodefense Strategy, Practice, and Science, 1*(3), pp. 203–214. doi:10.1089/153871303769201851

Carlson, R. (2008). *Gene synthesis cost update.* Retrieved March 18, 2009, from http://www.synthesis.cc/2008/11/gene-synthesis-cost-update.html

Caruthers, M. H., Beaucage, S. L., Becker, C., Efcavitch, J. W., Fisher, E. F., Galluppi, G., et al. (1983). Deoxyoligonucleotide synthesis via the phosphoramidite method. *Gene Amplification and Analysis, 3*, pp. 1–26.

Chan, L. Y., Kosuri, S., & Endy, D. (2005). Refactoring bacteriophage T7. *Molecular Systems Biology, 1*, p. 2005.0018. doi:10.1038/msb4100025

Dunn, J. J., & Studier, F. W. (1983). Complete nucleotide sequence of bacteriophage T7 DNA and the locations of T7 genetic elements. *Journal of Molecular Biology, 166*(4), pp. 477–535. doi:10.1016/S0022-2836(83)80282-4

Endy, D. (2005). Foundations for engineering biology. *Nature, 438*(7067), pp. 449–453. doi:10.1038/nature04342

Endy, D., You, L., Yin, J., & Molineux, I. J. (2000). Computation, prediction, and experimental tests of fitness for bacteriophage T7 mutants with permuted genomes. *Proceedings of the National Academy of Sciences of the United States of America, 97*(10), pp. 5375–5380. doi:10.1073/pnas.090101397

Integrated DNA Technologies (2005). *Chemical synthesis of oligonucleotides.* Retrieved March 18, 2009, from http://www.idtdna.com/TechVault/TechVault.aspx

Knight, T. (2003). *Idempotent vector design for standard assembly of biobricks.* Retrieved March 18, 2009, from http://dspace.mit.edu/handle/1721.1/21168

Knight, T. F. (2005). Engineering novel life. *Molecular Systems Biology, 1*, 2005.0020. doi:10.1038/msb4100028

Kosuri, S. (2007). *Simulation, models, and refactoring of bacteriophage T7.* Retrieved March 18, 2009, from http://hdl.handle.net/1721.1/35864

Kuldell, N. (2007). Authentic teaching and learning through synthetic biology. *Journal of Biological Engineering, 1*, 8. doi:10.1186/1754-1611-1-8

Petroski, H. (1985). *To engineer is human: The role of failure in successful design.* New York, NY: St. Martin's Press.

Registry of Standard Biological Parts (2009). Retrieved March 18, 2009, from http://partsregistry.org/Part:BBa_J45014

Stahler, P., Beier, M., Gao, X., & Hoheisel, J. D. (2006). Another side of genomics: Synthetic biology as a means for the exploitation of whole-genome sequence information. *Journal of Biotechnology, 124*(1), pp. 206–212. doi:10.1016/j.jbiotec.2005.12.011

Studier, F. W. (1969). The genetics and physiology of bacteriophage T7. *Virology, 39*(3), pp. 562–574. doi:10.1016/0042-6822(69)90104-4

Van Regenmortel, M. H. (2004). Reductionism and complexity in molecular biology. scientists now have the tools to unravel biological and overcome the limitations of reductionism. *EMBO Reports, 5*(11), pp. 1016–1020.

CHAPTER 2

Bacteriophage as Templates for Refactoring

2.1 INTRODUCTION

It is difficult to overstate the importance of bacteriophage to discovery science, to molecular biology research techniques, and to undergraduate teaching. The most recent compendium of bacteriophage information (Calendar, 2006) catalogs the impressive role of phage in the early development of molecular biology, the exquisite details known about every major phage family, and the numerous applications of phage to meet particular needs like pollution indicators and antimicrobial therapies. In the foreword to this authoritative compendium, however, is the recognition that interest has ebbed and flowed dramatically since the discovery of phage in 1915. As early as the 1930s, when phage gave disappointing results for control of microbial infections, the "glamour had begun to tarnish" (Cold Spring Harbor Laboratory of Quantitative Biology, Cairns, & Delbrück, 1966, p. 5). Then, successful use of phage to address mechanistic questions of inheritance led to a resurgence of interest in this experimental system. Research across many labs generated descriptions of phage life cycles, including that of lambda phage that was detailed in the readable and teacher-friendly book, *The Genetic Switch* (Ptashne, 2004). This publication established a lasting place for phage in undergraduate and graduate education. And yet by the 1970s, phage as a field of research had "all but died" (Calendar, 2006, Foreword). At the start of the 21st century, phage biology has reemerged, inspired by research in nanotechnology (Merzlyak & Lee, 2006), interest in bacterial evolution (Medhekar & Miller, 2007), and an appreciation of phage in control of bacterial pathogenesis (P. Garcia, Martinez, Obeso, & Rodriguez, 2008; Mattey & Spencer, 2008).

Redesign of bacteriophage genomes is a recent approach that has been applied to the T7 and the M13 phage (Chan, Kosuri, & Endy, 2005; Kuldell, 2007). Given that failure is often associated with early design efforts, the initial success in refactoring the T7 genome is all the more remarkable. Many factors contributed to the success of T7.1, including a considerable understanding of basic T7 biology on which the redesign was based, the relative independence of the T7 life cycle from that of the cell it infects, and the reasonably conservative approach taken by the researchers in the refactoring effort. Although comparable systems-level and parts-level understanding exists for the other bacteriophage that has been reengineered, namely, M13, the refactoring and retasking of M13 met with less success. As detailed later in this chapter, even subtle modifications to the M13 genome had profound consequences to phage viability, making the comparison of T7 and M13 biology and refactoring instructive for future genome engineering efforts.

2.2   T7 BACKGROUND

As early as the 1930s, Max Delbruck recognized bacteriophage as a "gadget" worthy of study to illuminate key genetic questions (Calendar, 2006, p. 4). Investigations of phage biology by all-star scientists like Delbruck, Salvador Luria, Alfred Hershey, and many others addressed critically important mechanistic questions of heredity (e.g., the nature and cause of gene mutations) and defined the methods for studying these questions. Profoundly successful in spearheading this line of inquiry, Delbruck was said to have "provided the ideological and spiritual fountainhead for the discipline that would eventually style itself molecular genetics" (Cold Spring Harbor Laboratory of Quantitative Biology, et al., 1966, p. 352).

   The T-phages (T for type) were numbered 1 through 7 and from the outset of phage research served as the model to study the lytic phage life cycle (infections with obligate lysing of the bacterial host during production of new phage particles). Lysis of the bacterial host could be visualized by a resulting "plaque" or clearing in an otherwise opaque bacterial lawn (Figure 2.1). Although all the T-phages could grow and lyse *Escherichia coli*, the T7 phage soon became the prototypical member of the T-phage family. This was due in part to practical considerations: T7 forms large plaques, and because the phages die quickly upon drying, they do not persistently contaminate the laboratory (Calendar, 2006, p. 277).

   However, the scientific importance of T7 among its T-phage cousins was established in 1969 by F. William Studier, who identified 19 genes of the T7 genome by analyzing a collection of amber mutations in the phage (Studier, 1969). The characterization of these mutants included elegant genetic mapping studies, which showed clustering of related functions, and complementation analysis, for example, grouping the "leftmost" genes 1–6 for their defects in DNA synthesis and host lysis. As noted by the author in this important publication, the simple life cycle of the phage and the associated genetic tools to investigate it made T7 particularly attractive for study. Within a handful of years, an additional two genes on the T7 genome were characterized genetically (Masamune, Frenkel, & Richardson, 1971), and another few were identified biochemically (Studier & Maizel, 1969; Studier, 1969).

   Thus, with time and effort, the fraction of the T7 genome deemed physiologically meaningful increased incrementally. However, it was the sequencing of the full genome that brought the next

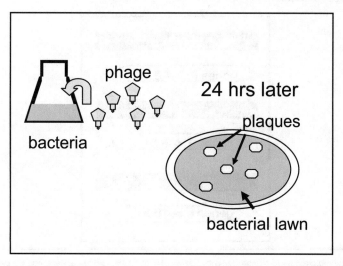

FIGURE 2.1: Plaque assay.

leap in identifiable functional elements of the genome (Dunn & Studier, 1983). All the previously identified genes were found within the 39,936 nucleotides of the T7 sequence. In addition, several "unsuspected genes" were identified, raising the predicted number of T7 genes from 38 to 50. The full nucleotide sequence also elucidated the compact architecture of the genome and suggested that all 50 genes should be expressed.

The current molecular description of the natural T7 bacteriophage details many genomic features and their role in the phage life cycle (Figure 2.2). T7 infects of its *E. coli* host with a double-stranded DNA genome of approximately 40 kilobases encoding 56 known or potential genes (Calendar, 2006, p. 279). Terminal repeats bookend the genome while genes and regulatory elements are tightly packed within it. Genetic elements show short regions of overlap that will be detailed below. All but 8% of the genome is believed to encode proteins (Calendar, 2006, p. 281). T7 genes have been named according to their position from "left" to "right" on the genome. Thus, gene 0.3 maps to the leftmost portion of the linear genome and gene 19 to the rightmost section. Originally, whole numbers were used to signify essential genes and non-integers designated nonessential genes (Calendar, 2006; Studier, 1969). With further study of the T7 phage, this nomenclature broke down, but the legacy naming remains.

The mechanisms of T7 infection are also well understood. The infection cycle shows three bursts of phage protein synthesis. The early genes are transcribed by the host's RNA polymerase machinery and then translated within 8 minutes of infection. Most of these early proteins help

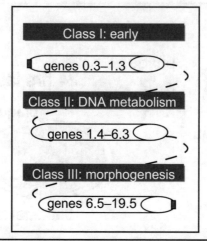

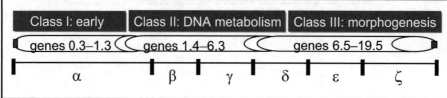

FIGURE 2.2: T7 genome organization.

establish a cellular environment that favors phage infection. All the early genes are nonessential to the phage except gene 1, which encodes the phage's own RNA polymerase enzyme. Phage genes relevant later in the infection cycle are transcribed with this T7 RNA polymerase. Indeed, efficient entry of the whole genome into the *E. coli* host seems to require the T7 RNA polymerase to "pull" the intermediate and late genes though a phage-made pore (L.R. Garcia & Molineux, 1995; L.R. Garcia & Molineux, 1996). These intermediate and late genes control T7 phage DNA replication and the packaging of new phage virions. These processes depend on materials provided by the host cell (e.g., dNTPs, amino acids) but occur independent of host replication.

Despite the detailed appreciation of the phage genome and its infection processes, many questions remain unresolved. How, for example, can we reconcile the fact that there is no direct evidence for several of the proteins predicted by the sequence analysis? Traditional scientific approaches to further understand the T7 genome might include sequence comparison with that of other T-family phage. In parallel, deletion analysis of the uncharacterized genes (individually and in combination) could be applied with phenotypic profiling of the resulting constructs. Alternatively, the genes might be ignored, and only data for characterized genes might be applied to computer simulations of the T7 life cycle. However, computer models built this way do not adequately or even accurately predict T7 behavior (Endy, Kong, & Yin, 1997). In 2005, an alternative approach was described, that of genome refactoring (Chan et al., 2005).

2.3   REBUILDING T7

The goals for reengineering the T7 genome were explicit and thoughtful. Although motivated by a desire to make a more "modelable" system, the design criteria are a helpful framework in considering any genome-engineering endeavor. The design goals included the parsing of the genome into definable functions, the "unstuffing" of functions so each part was encoded by a sequence that was disentangled from the sequences for other parts, and the limiting of each part to encode only one function (although tightly linked functions, like an RBS and its gene, were annotated as a single part). In addition, parts were flanked so as to facilitate future genome manipulations. The design goals were balanced by the researchers' appreciation that the changes they called for had to be "buildable" with existing DNA synthesis technologies.

At the outset, it was uncertain how many simultaneous changes the phage could tolerate. Unlike the M13 phage, detailed below, only viable T7 phage can be recovered, amplified, and propagated. Taking a conservative approach to redesign, the investigators divided the genome into subsections. These subdivisions made the construction and testing process more manageable. Thus, the initial design for T7.1 could be limited to the leftmost section of the genome, replacing the natural 11,515 bases with 12,179 bases of designed sequences. The boundary for the T7 refactoring was guided by the presence of unique restriction sites in the natural T7 genome, restriction sites that bracket the leftmost sections of sequence. Two subsections, designated "alpha" and "beta," were built and combined with the remaining portions of the natural genome.

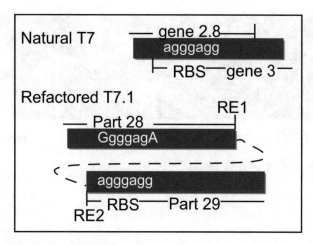

FIGURE 2.3: T7.1 division of genes 2.8 and 3.

(A) as built vs. as designed

Location on T7.1 (Genome Position)	Nature of Difference	Probable Reason for Difference	Expected Outcome
DIL (164-170), DIR & D2L (338-350)	Restriction sites were not added in construction	Difficulties in manipulating left end of genome resulted in using wild-type	Loss of manipulability in part 1 (containing A0)
gene *0.4* (1418)	Single base deletion	Unknown	Frameshift after 27th amino acid followed by early termination of gene *0.4*
D0L (1304-1310) D0R (1494-1500)	Restriction sites appear twice	Inefficiency of digestion of scaffold	No expected change
gene *0.6B*	Single base addition	Error is known to be in stock of wild-type genome	Dependent upon nature of putative translational slippage in formation of gene *0.6B*
D11L (3302-3307)	Restriction site appears twice	Inefficiency of digestion of scaffold	No expected change
gene *1* (4877)	Single base mutation	Error in PCR or within wild-type genome	Silent mutation, no expected change
gene *1* (5159)	Single base mutation	Error in PCR or within wild-type genome	Silent mutation, no expected change
gene *1* (5399)	Single base mutation	Error in PCR or within wild-type genome	Silent mutation, no expected change
D14R (6591-6597)	Restriction site appears twice	Inefficiency of digestion of scaffold	No expected change
TE (7827)	Single base deletion	Primer synthesis error	Possible loss of function of transcriptional terminator
D20L (8082-8086)	Restriction site appears twice	Inefficiency of digestion of scaffold	No expected change
U4 (8153-8159)	Restriction site was not added in construction	Failure in site-directed mutagenesis	Loss of manipulability of overlap in parts 18 and 19
D22L (8247-8253)	Restriction site appears twice	Inefficiency of digestion of scaffold	No expected change

(B) plaque assay for T7.1 rebuild

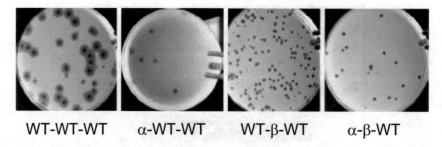

WT-WT-WT   α-WT-WT   WT-β-WT   α-β-WT

FIGURE 2.4: Results of T7.1 rebuild. (from Chan et al., 2005).

Within the alpha and beta sections of the T7 genome were 32 of the 73 recognized biological "parts." Many of the parts were seen to have overlapping physical boundaries. For example, the RBS used to initiate translation from the gene 3 transcript lies within the coding sequence for gene 2.8 (Figure 2.3). This genomic architecture makes the independent manipulation of gene 2.8 and gene 3 impossible. The researchers refactored this section of the genome by introducing silent point mutations into gene 2.8 to render the RBS of gene 3 nonfunctional. Next, the sequence overlaps between gene 2.8 and gene 3 were disentangled by codon shuffling in each open reading frame to limit direct sequence repeats. Such direct repeats were known to be hot spots for recombination, and the researchers went to considerable efforts throughout their redesign to limit the introduction of repeats into the refactored genome. Finally, the modular parts from gene 2.8 and gene 3 were renamed parts 28 and 29 and bracketed by restriction sites *Bam*HI and *Eag*I, respectively.

After proceeding through the alpha and beta sections of the T7 genome in this fashion, the refactored segments were recombined individually and in combination with the remainder of the natural genome. A few cases were noted where the resulting "build" differed from the sequence that was specified. For example, part 1 could not be excised from the refactored genomes as designed (Figure 2.4). Sequencing of the "as-built" genome revealed some single-base deletions versus the specified sequence. Most remarkable, however, was the observation that the three variants, namely, the alpha-WT-WT, the WT-beta-WT, and the alpha-beta-WT genomes, only modestly changed the lysis kinetics of the host by the phage. This property was assessed through growth curves of infected cells as well as by plaque assay. Plaques were found to be smaller for the refactored T7, but were still formed, indicating that even in the absence of complete understanding of the natural genomic elements, a partial refactoring of the genome could be thoughtfully performed and constructed.

The viability of T7.1 provides an instance proof for the successful redesign and rebuilding of genomes. We now discuss subsequent attempts to apply these insights to the refactoring of a different bacteriophage—M13.

2.4 M13 BACKGROUND

The T-family of phage is only one subtype of bacteriophage, one of more than 5000 distinct subtypes, making bacteriophage the largest viral group found in nature (Ackermann, 2007). A family of bacteriophages quite distinct from the T-family (according to their morphology and nucleic acid content) is the filamentous phage family, of which M13 is the best studied. M13's genome exists as single-stranded circular DNA when it is packaged in the phage filament but is replicated to a double-stranded circular genome when it infects its *E. coli* host. M13 is a nonlytic phage, allowing it to establish a persistent infection of its *E. coli* host (Figure 2.5). Although an infected host can shed approximately 1000 phage/hour (Calendar, 2006, p. 150), the phage does not considerably impact the cell's physiology.

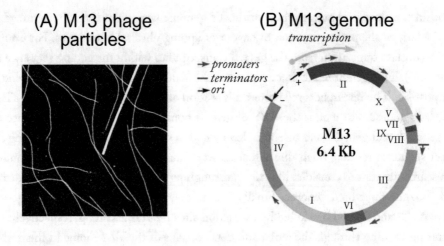

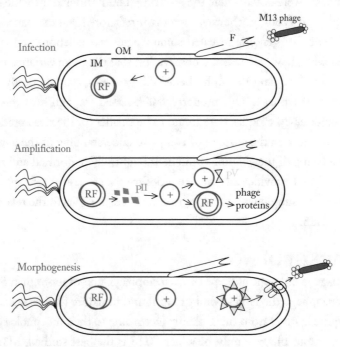

FIGURE 2.5: (A) EM of M13 phage particles from M. Simon. (B) M13 genome, image from M. Blaber. (C) Schematic of M13 life cycle.

The M13 virion has a long (~900 nm), narrow (~20 nm) protein coat that encases a small (~6.4 kb) genome (Figure 2.5). The genome encodes 11 proteins, 5 of which are exposed on the phage's protein coat and six of which are involved in phage maturation inside its *E. coli* host. Proteins and genes are correspondingly numbered, for example, gene 3 encodes protein 3 (p3). M13 phage and all its filamentous phage cousins infect *E. coli* through a bacterial structure known as the F pilus, with the M13 coat protein called p3 contacting the TolA protein on the bacterial pilus. The phage genome is then transferred through the pilus to the cytoplasm of the bacterial cell where resident proteins convert the single-stranded DNA genome to a double-stranded replicative form. This DNA then serves as a template for expression of the phage genes. Amplification of the phage genome as single-stranded DNA for packaging requires the phage-encoded proteins p2 and p10 as well as several host-encoded enzymes and raw materials. As detailed below, the natural M13 genome has p10 fully embedded within the open reading frame for p2.

Once single-stranded DNA has been amplified, the DNA is extruded from the cell through a phage-encoded pore, with five phage proteins encasing the genome. The major coat protein is a 50-amino-acid protein called p8. It takes approximately 2700 copies of p8 to make the approximately 900-nm-long phage capsid that encases a natural M13 genome. The capsid's dimensions are flexible though and the number of p8 copies adjusts to accommodate the size of the single-stranded genome it packages. For example, when the phage genome was mutated to reduce its number of DNA bases (from 6.4 kb to 221 bp) (Specthrie, et al., 1992), the p8 coat "shrink wrapped" around the reduced genome, decreasing the number of p8 copies to less than 100. Viable phage with increased DNA content seem to top out at approximately twice the natural phage virion length, an indication that there is some but not unlimited plasticity in this genetic system.

Four other proteins are presented on the M13 capsid with p8. At one end of the capsid are five copies of the surface exposed p9 and a more buried companion protein, p7. If p8 forms the shaft of the phage, p9 and p7 form the "blunt" end that is seen in electron micrographs. These proteins are some of the smallest known (only 33 and 32 amino acids), although some additional residues can be added to the N-terminal portion of each, which are then presented on the "outside" of the phage coat (Gregori et al., 2000). At the other end of the phage particle are five copies of the surface exposed p3, already described for its role in contacting the bacterial pilus during infection, and a less exposed accessory protein, p6. When new phage bud from the bacterial surface, p3 is the last point of contact between the phage particle and the host cell. Some deletions in p3 prevent full escape from the host, and phages that are 10–20 times the normal length and with several copies of the phage genome are produced (Weiss, Roth, Baldi, & Sidhu, 2003).

Two properties of M13's life cycle, namely, its natural conversion between single-stranded and double-stranded DNA and its naturally balanced coexistence with its infected host, have made M13 a powerful tool for molecular biology (Vieira & Messing, 1987). In particular, M13-based tools for DNA sequencing and DNA mutagenesis were widespread before the availability of automated DNA sequencing machines and readily available mutagenesis kits. Although other experimental tools have since replaced M13-based sequencing and mutagenesis, several ingenious modifications were made to the phage genome to facilitate the experimental work. These modifications included the addition of a drug-selectable marker and a bacterial origin of replication to the phage genome, as well as the introduction of multiple cloning sequences. The modified phage, designated M13KO7, has been used as a "helper phage" because it replicates its own genome relatively slowly, allowing DNA from an alternative, more natural copy of M13 to be preferentially packaged into the virion.

These modifications to the natural genome originally supported M13 as a tool for molecular biology, but the M13KO7 variant is also an ideal phage genome to refactor. First, the M13KO7 variant does encode a viable phage that can be titered with a plaque assay on specialized bacterial strains. Second, and more important, any changes made to the M13KO7 DNA that disable the phage infection cycle can still be propagated as "phagmid" DNA using standard transformation and selection. The T7 and M13 refactoring are critically different in this regard. T7.1 could only be recovered if the simultaneous changes made to the genome supported the phage as an infectious agent.

Another M13-based research tool, still widely used and relevant, is phage display of peptides and proteins. In phage display, foreign sequences that are immunologically, enzymatically, or pharmacologically relevant can be discovered by fusing a collection of candidate sequences to a phage coat protein. The pool of modified M13 phage can then be affinity purified for the desired activity. The first example of phage display (Smith, 1985) demonstrated the utility of this technique by fusing a portion of the *Eco*RI restriction endonuclease to the phage "tail" protein called p3. A unique *Bam*HI site in the gene for p3 allowed a fragment of the *Eco*RI sequence to be inserted in frame, and the resulting phage particle retained infectivity, albeit approximately 100-fold lower than normal, and displayed the *Eco*RI sequence on the phage surface in an accessible form.

The usefulness of phage display was recognized immediately, and several modifications to the genome were engineered (for review, see Sidhu, Feld, & Weiss, 2007). For example, two new unique restriction sites, *Kpn*I and *Eag*I, were introduced in place of the unique *Bam*HI site in the gene for p3 (*Ph.D. peptide display cloning system (E8101), phage display, NEB*). This modification enabled users of phage display to orient fragments in the M13 genome rather than allowing fusions to insert either "forward" or "backward" as was the case originally. Currently, most phage display uses only two of the five proteins on M13's coat, namely, p3 and p8. Although p8 is orders of magnitude more abundant on the phage particle's surface than p3, the protein is more constrained in terms of the size of the peptide it can present. This is because the thousands of copies of p8 must align into a semicrystalline array as they construct the filamentous shaft of the phage coat. Thus, the p8 protein tolerates only short peptide fusions, presumably because larger, bulkier additions disrupt the phage packaging (Clackson & Lowman, 2004; Kishchenko, Batliwala, & Makowski, 1994).

The compact nature of the M13 genome has hampered the deployment of other phage proteins for display. For example, the start codon for the p8 is upstream of the stop codon for another phage coat protein, p9, making their independent control and modification difficult. With a refactored M13 genome, protein fusions to p9 could be engineered without affecting p8, for example. Taken a step further, a genome optimized by humans rather than nature might have multiple options for directed cloning into every phage coat protein.

2.5   REBUILDING M13

Although following the success of T7 refactoring, the M13 redesign was motivated by different goals. First, we were motivated by the usefulness of refactoring as an educational framework. The M13 refactoring was situated within an undergraduate laboratory class for biological engineers at MIT. The protocols that our students followed made explicit distinction between small changes to the genome with established techniques of genetic engineering and the whole-cloth rewriting of the genome to meet discovery and application end points. Second, the modification of the M13 genome was motivated by the renewed interest in this phage as a template for nanowires and for biomaterials. These motivations led to numerous, varied designs for the M13 phage genome, as will be detailed in Chapter 3.

Like the "unstuffing" of the T7 genes 2.8 and 3, there are easy-to-spot elements in the M13 sequence that are natural candidates for initial refactoring efforts, particularly genes 2 and 10, and genes 8 and 9 (Figure 2.6).

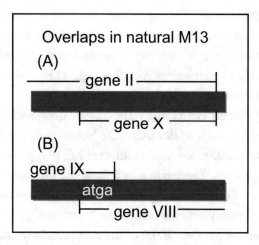

FIGURE 2.6: M13.1 figure genes 2 and 10, and genes 9 and 8.

Genes 2 and 10 are in-frame, fully overlapping genes. The two phage gene products play a critical role in the amplification of the M13 genome. p2 nicks the double-stranded form of the genome to initiate replication of the single strand destined for packaging. Without p2, replication of the phage genome cannot occur. Another phage-encoded protein, p10, is important for regulating the number of double-stranded genomes in the bacterial host. What is particularly interesting about p10 is that it is identical to the C-terminal portion of p2 because the gene for p10 is embedded in the gene for p2, and the protein arises from transcription initiation within gene 2. Interestingly, it is the translation of p2 that regulates expression of p10 because elongating ribosomes translating the p2 message appear to interfere with internal initiation from the p10 RBS (Yu, Kokoska, Khemici, & Steege, 2007). This complex but fascinating arrangement precludes the possibility for recombination between the genes and makes for a compact and efficient natural genome. However, it also makes the manipulation of p10 inextricably linked to the manipulation of p2, a potential engineering headache.

A second spot on the M13 genome seemingly ripe for refactoring is the 5′ end of gene 8 that is entangled with the 3′ end of gene 9. This arrangement is reminiscent of the natural T7 arrangement for genes 2.8 and 3, which were successfully disengaged in the T7.1 phage. The 1980 article (van Wezenbeek, Hulsebos, & Schoenmakers, 1980) that reported the full genomic sequence of M13 identified gene 9 as encoded from bases 1206 to 1304 on the circular genome. Given the expected small size of a protein translated from gene 9 (only 32 amino acids) and the absence of lethal mutations that mapped to this region of the genome, researchers had originally presumed gene 9 to be noncoding "leader" sequence for gene 8 (Sugimoto, Sugisaki, Okamoto, & Takanami, 1977). Understanding of the natural system was significantly advanced by the confirmation of the open reading frame for gene 9 (van Wezenbeek et al., 1980) as well as the demonstration of the protein, p9, on the M13 phage coat (Simons, Konings, & Schoenmakers, 1979). Nonetheless, an understanding of the natural architecture did not make the independent manipulation of genes 9 and 8 any more trivial. Indeed, it was clear that any sequence fused to the N-terminus of gene 8 would, by necessity, add sequence to the C-terminus of gene 9 as well. The peptide fused to p9 would be out of frame relative to the same sequence added in frame to gene 8.

The "helper phage" system can partially circumvent this problematic organization of the genome. With a helper phage, such as the M13KO7 helper phage, a second M13-like plasmid (termed a *phagemid*) can be transformed into an infected cell to express only the modified p8 and none of the other phage proteins. The modified p8 from the phagemid and the natural p8 from the M13KO7 helper genome both can incorporate into the phage coat but at a relative ratio that is almost uncontrollable and is nearly unpredictable. If an N-terminal p8 fusion could be made to the M13 genome without affecting p9, then 100% of the coat protein would bear this fusion. A successfully refactored genome, with genes 9 and 8 physically separated as well as bracketed by unique restriction sites, would enable this manipulation. This goal, among others, was addressed with the M13 refactoring project.

2.6   REFLECTION ON M13 REFACTORING

In many instances the goals of research and the goals of teaching are at odds. Unpredictable outcomes, unexpected variations, and surprising findings are among the most exciting parts of scientific research because they can lead to new questions as well as refined understanding of the question at hand. In a classroom setting, however, such floundering and searching can derail learning. Conventional wisdom argues that students best find their own footing when some of the complexity and noise in the information are massaged away. It is ironic that a field like synthetic biology, which promotes the managing of complexity in biological systems, is also positioned to upend the conventional wisdom and teach by celebrating unexpected discoveries and findings. As the next chapter details, our goals in reengineering M13 were met not in a traditional research laboratory setting but rather in the context of an undergraduate class for biological engineers at MIT.

REFERENCES

Ackermann, H. W. (2007). 5500 phages examined in the electron microscope. *Archives of Virology*, *152*(2), pp. 227–243. doi:10.1007/s00705-006-0849-1

Calendar, R. (Ed.). (2006). *The bacteriophages* (2nd ed.). Oxford: Oxford University Press. doi:10.1086/509419

Chan, L. Y., Kosuri, S., & Endy, D. (2005). Refactoring bacteriophage T7. *Molecular Systems Biology*, *1*, p. 2005.0018. doi:10.1038/msb4100025

Clackson, T., & Lowman, H. B. (Eds.). (2004). *Phage display: A practical approach*. Oxford: Oxford University Press.

Cold Spring Harbor Laboratory of Quantitative Biology, Cairns, J., & Delbrück, M. (1966). *Phage and the origins of molecular biology [essays]*. Cold Springs Harbor, New York. CSHL Press

Dunn, J. J., & Studier, F. W. (1983). Complete nucleotide sequence of bacteriophage T7 DNA and the locations of T7 genetic elements. *Journal of Molecular Biology, 166*(4), pp. 477–535. doi:10.1016/S0022-2836(83)80282-4

Endy, D., Kong, D., & Yin, J. (1997). Intracellular kinetics of a growing virus: A genetically structured simulation for bacteriophage T7. *Biotechnology and Bioengineering, 55*(2), pp. 375–389. doi:10.1002/(SICI)1097-0290(19970720)55:2<375::AID-BIT15>3.0.CO;2-G

Garcia, L. R., & Molineux, I. J. (1995). Incomplete entry of bacteriophage T7 DNA into F plasmid-containing *Escherichia coli. Journal of Bacteriology, 177*(14), pp. 4077–4083.

Garcia, L. R., & Molineux, I. J. (1996). Transcription-independent DNA translocation of bacteriophage T7 DNA into *Escherichia coli. Journal of Bacteriology, 178*(23), pp. 6921–6929.

Garcia, P., Martinez, B., Obeso, J. M., & Rodriguez, A. (2008). Bacteriophages and their application in food safety. *Letters in Applied Microbiology, 47*(6), pp. 479–485. doi:10.1111/j.1472-765X.2008.02458.x

Gregori, S., Bono, E., Gallazzi, F., Hammer, J., Harrison, L. C., & Adorini, L. (2000). The motif for peptide binding to the insulin-dependent diabetes mellitus-associated class II MHC molecule I-Ag7 validated by phage display library. *International Immunology, 12*(4), pp. 493–503. doi:10.1093/intimm/12.4.493

Kishchenko, G., Batliwala, H., & Makowski, L. (1994). Structure of a foreign peptide displayed on the surface of bacteriophage M13. *Journal of Molecular Biology, 241*(2), pp. 208–213. doi:10.1006/jmbi.1994.1489

Kuldell, N. (2007). Authentic teaching and learning through synthetic biology. *Journal of Biological Engineering, 1*, p. 8. doi:10.1186/1754-1611-1-8

Masamune, Y., Frenkel, G. D., & Richardson, C. C. (1971). A mutant of bacteriophage T7 deficient in polynucleotide ligase. *The Journal of Biological Chemistry, 246*(22), pp. 6874–6879.

Mattey, M., & Spencer, J. (2008). Bacteriophage therapy—cooked goose or phoenix rising? *Current Opinion in Biotechnology, 19*(6), pp. 608–612. doi:10.1016/j.copbio.2008.09.001

Medhekar, B., & Miller, J. F. (2007). Diversity-generating retroelements. *Current Opinion in Microbiology, 10*(4), pp. 388–395. doi:10.1016/j.mib.2007.06.004

Merzlyak, A., & Lee, S. W. (2006). Phage as templates for hybrid materials and mediators for nanomaterial synthesis. *Current Opinion in Chemical Biology, 10*(3), pp. 246–252. doi:10.1016/j.cbpa.2006.04.008

Ph.D. peptide display cloning system (E8101), phage display, NEB. Retrieved March 19, 2009, from http://www.neb.com/nebecomm/products/productE8101.asp.

Ptashne, M. (2004). *A genetic switch: Phage lambda revisited* (3rd ed.). Cold Spring Harbor, NY: Cold Spring Harbor Laboratory Press.

Sidhu, S. S., Feld, B. K., & Weiss, G. A. (2007). M13 bacteriophage coat proteins engineered for improved phage display. *Methods in Molecular Biology (Clifton, N.J.), 352*, pp. 205–219. doi:10.1385/1-59745-187-8:205

Simons, G. F., Konings, R. N., & Schoenmakers, J. G. (1979). Identification of two new capsid proteins in bacteriophage M13. *FEBS Letters, 106*(1), pp. 8–12. doi:10.1016/0014-5793(79)80683-3

Smith, G. P. (1985). Filamentous fusion phage: Novel expression vectors that display cloned antigens on the virion surface. *Science (New York, NY), 228*(4705), pp. 1315–1317. doi:10.1126/science.4001944

Specthrie, L., Bullitt, E., Horiuchi, K., Model, P., Russel, M., & Makowski, L. (1992). Construction of a microphage variant of filamentous bacteriophage. *Journal of Molecular Biology, 228*(3), pp. 720–724. doi:10.1016/0022-2836(92)90858-H

Studier, F. W. (1969). The genetics and physiology of bacteriophage T7. *Virology, 39*(3), pp. 562–574. doi:10.1016/0042-6822(69)90104-4

Studier, F. W., & Maizel, J. V., Jr. (1969). T7-directed protein synthesis. *Virology, 39*(3), pp. 575–586. doi:10.1016/0042-6822(69)90105-6

Sugimoto, K., Sugisaki, H., Okamoto, T., & Takanami, M. (1977). Studies on bacteriophage fd DNA. IV. the sequence of messenger RNA for the major coat protein gene. *Journal of Molecular Biology, 111*(4), pp. 487–507. doi:10.1016/S0022-2836(77)80065-X

van Wezenbeek, P. M., Hulsebos, T. J., & Schoenmakers, J. G. (1980). Nucleotide sequence of the filamentous bacteriophage M13 DNA genome: Comparison with phage fd. *Gene, 11*(1–2), pp. 129–148.

Vieira, J., & Messing, J. (1987). Production of single-stranded plasmid DNA. *Methods in Enzymology, 153*, pp. 3-11. doi:10.1016/0076-6879(87)53044-0

Weiss, G. A., Roth, T. A., Baldi, P. F., & Sidhu, S. S. (2003). Comprehensive mutagenesis of the C-terminal domain of the M13 gene-3 minor coat protein: The requirements for assembly into the bacteriophage particle. *Journal of Molecular Biology, 332*(4), pp. 777–782. doi:10.1016/S0022-2836(03)00950-1

Yu, J. S., Kokoska, R. J., Khemici, V., & Steege, D. A. (2007). In-frame overlapping genes: The challenges for regulating gene expression. *Molecular Microbiology, 63*(4), pp. 1158–1172. doi:10.1111/j.1365-2958.2006.05572.x

CHAPTER 3

Methods/Teaching Protocols for M13 Reengineering

3.1 INTRODUCTION

Reengineering the M13 genome offers rich opportunities to contribute to the field of synthetic biology and to biological engineering more broadly. In addition, this project offers opportunities for the next generation of biological engineers to gain hands-on experience in the practices of authentic inquiry. In other words, rather than undergraduate laboratory experiences in which students focus primarily on learning techniques and "rediscovering" known phenomena, a focus on reengineering M13 engages students in the authentic practices of professionals, albeit in the relatively controlled conditions of an academic research laboratory scaled up for 25 to 35 neophyte researchers exploring a common topic.

Motivated by this desire to engage students in authentic practices, we decided to pursue M13 reengineering in the context of a laboratory class required of all biological engineering majors at MIT. This laboratory project not only taught technical skills and scientific content but also exposed students to the practice of engineering as applied to the complex living world. The "sage on stage" model for teaching, in which a wise professor lectures to a sea of fully naive students, was not a possibility because the M13 bacteriophage, although extensively studied, had never been refactored nor had its genome been so extensively reengineered. Students who engaged in this series of experiments could not merely be proficient mimics of experienced bioengineers. Instead, they had to bring their own ideas to bear on the problem, deciding what is most useful to do and then studying the intended and unintended consequences of their design choices. As described in the next chapter, our students attest to the value of learning biological engineering by doing some.

A second valued lesson our students take away from the M13 reengineering project is the importance of "to the root" engineering (Lima, 2007). There will always be a place for engineers who can apply existing knowledge to make something useful and an even greater place for engineers who can add back to the information base itself through their work. What has been only recently appreciated is the educational imperative to produce engineers who also consider and address the root causes from which the engineering need arose. It will be the job of future engineers to wisely solve a problem at hand as well as to remediate the imbalance, ignorance, or inequity that gave rise to it. The strongest recent example of the need to train engineers this way is seen in the flooding of New Orleans by Hurricane Katrina in 2005. This natural disaster was made truly catastrophic by the failure of the levees that resulted in flooding of 80% of the city. The levees were long known to be predictable failures, with consequences compounded by the ongoing destruction of wetlands and land subsidence. The hurricane's destruction of New Orleans left many senior engineers wondering how they might have strengthened their positions with respect to setting a sensible agenda and to allocating resources that support the broadest community. Engineering education is among the factors being reconsidered in this vein, including reconsideration of college-level curricula so the next generation of engineers can better appreciate both short-term and long-term challenges they will face (Lima, 2007; see also similar reconsiderations after the Challenger and Columbia space shuttle disasters, e.g., Browning, 1988; Guthrie & Shayo, 2005). Such improved curricula require we teach both technical and communication skills that empower students to first envision and then to lobby for an appropriate, far-sighted solution.

The issues raised by the reengineering of the M13 genome are not on the order of those raised by the flooding and rebuilding of New Orleans. Nevertheless, M13 does provide a genomic template that for years has been tweaked and fiddled with to extend its application for phage display, site-directed mutagenesis, and sequencing. The newest generation of biological engineers should understand and gain proficiency in genetically engineering M13 through such ad hoc methods. However, they should also appreciate the benefits and challenges associated with restructuring the genome in more sweeping strokes. Lessons to teach both minor retooling and "to the root" engineering are described in the following sections.

3.2   GOALS AND OUTCOMES FOR "SIMPLE" GENETIC ENGINEERING OF M13

To ground the students in traditional methods for genetically engineering portions of the existing M13KO7 genome, we tasked them with a reasonably simple design challenge, namely, to add two unique restriction sites to the 3' end of the gene for p3. This modification of the genome was motivated with a pragmatic desire to engineer a better phage display system that supported directed cloning to the p3 coat protein. As described in Chapter 2, the ability to directionally insert DNA into the gene for any of the phage coat proteins is highly desirable and improves the efficiency of this research tool. Students were also encouraged to consider this effort as a means for making discoveries. The changes they applied to the M13KO7 phage genome were new, so any results, either positive or negative, would offer information about the phage's tolerance for manipulation. Any genomes with tolerated modifications were then available as a tool to discover new things about how the phage is organized and operates.

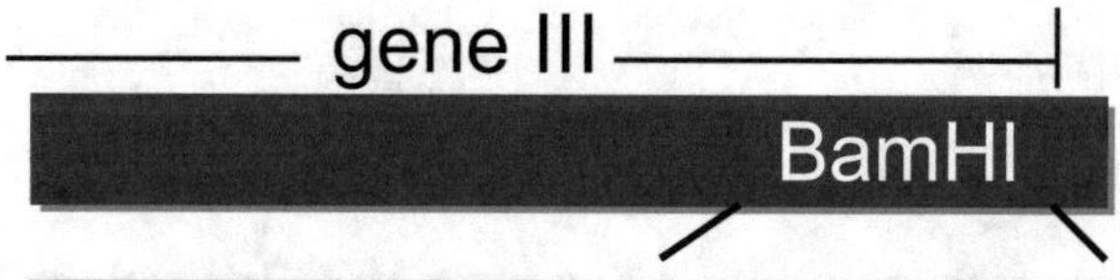

Oligo (top of pair)	Added Restriction Sites
5' GATCTGCGCACGCGT	FspI and MluI
5' GATCTAGACGATATC	XbaI and EcoRV
5' GATCTCGACGGCGCC	SalI and SfoI
5' GATCAGATCTTAGGCG CCCGA	BglII and SfoI
5' GATCTGATCAGTACT	BclI and ScaI
5' GATCTTAAGGACGTC	AflII and ZraI
5' GATCGGTACCACGTC	KpnI and BmgI
5' GATCATGAATTCGAT ATC	EcoRI and EcoRV
5'GATCAGATCTCCCACGTC	BglII and BmgBI
5' GATCGAATTCACGTC	EcoRI and BmgBI
5'GATCGAATTCTTGTTAAC	EcoRI and HpaI

FIGURE 3.1: Final designs for 11 oligonucleotides.

The traditional tools of molecular biology were sufficient to support the prescribed modifications to M13KO7. Students used the sequence data that are available in public databases to identify a natural restriction site, *Bam*HI, in the gene for p3 (van Wezenbeek, Hulsebos, & Schoenmakers, 1980; *Ph.D. peptide display cloning system (E8101), phage display, NEB*) into which they could insert their desired sequence additions. Their selected additions to the genome were fabricated through the synthesis of custom oligonucleotides. Students specified a pair of single-stranded DNA sequences that, when annealed, would provide recognition sites for two new restriction enzymes flanked by single-stranded overhangs. These overhanging sequences were complementary to the unique *Bam*HI site in the gene for p3, allowing for insertion of the annealed pair of oligonucleotides into the existing genome.

Although the students were reassured that there was no single "right" answer for the sequences they could insert, they also appreciated how several technical considerations made some choices wiser than others. Any DNA sequence they added to the gene for p3 gets translated to

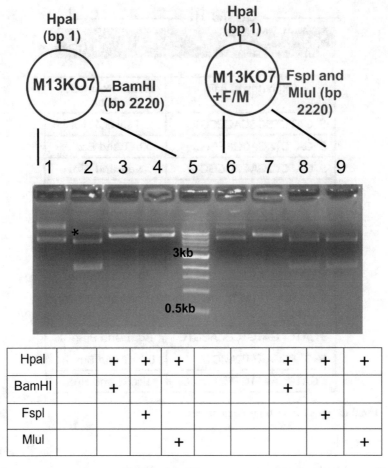

FIGURE 3.2: Genetic engineering of M13 gene III.

amino acids within the protein. Thus, beyond confirming the correct reading frame for the insertion, the students translated the resulting sequence and modified their design if translation gave rise to "undesirable" amino acids, such as proline. They also considered codon usage tables because any rare codons they inadvertently included might affect the expression level of p3 from their engineered genome. The final designs for 11 oligonucleotides are shown in Figure 3.1.

Multiple methods were used to assess insertion of the desired sequence and to analyze of the resulting phage. Diagnostic digests with agarose gel electrophoresis were used to identify candidates with insertions in the gene for p3, and sequencing was carried out to determine the orientation of the insert. The consequences of the insertions on phage properties were determined by Western analysis to evaluate p3 concentrations. Plaque assays were used to assess any consequences that

TABLE 1: Phage titers for gene III engineered phage.			
PHAGE	bp 2220	TITER (pfu/ml)	PLAQUE APPEARANCE
M13KO7	*Bam*HI	10^{14}	Normal
NB276	*Fsp*I/*Mlu*I	10^{14}	Normal
NB286	*Xba*I/*Eco*RV	10^{14}	Normal

the insertions had on phage propagation. Of the 13 designs that were specified and assembled (*Talk:20.109(F07):Start-up genome engineering—OpenWetWare*), only two clones were verified to carry the insert in the genome, and only one of those two carried the insert in the intended orientation. Both modified genomes generated phage titers comparable to those of the original M13KO7. The reasons for the failure of the other cloning candidates were not explored but could include the limited proficiency of the students themselves with the techniques of genetic engineering, a defect in the materials provided (e.g., errors in the oligonucleotides), and an inability of the phage to tolerate even seemingly benign changes. Figure 3.1 and Table 1 present the data collected for the two phages that were successfully modified.

What lessons did the students learn from this experience of genetically engineering a seemingly simple modification to the genome of M13KO7? Scientifically speaking, the complexity of living systems was made clear from the limited success of what initially appeared to be a small perturbation to the genome. From an engineering perspective, students gained first-hand experience with the practice of failure analysis and redesign based on the behaviors of prototypes. Most lasting, however, was the impact of this project in conjunction with the M13 refactoring project on the students' identity as biological engineers, outcomes that are further detailed in Chapter 4.

3.3   GOALS AND OUTCOMES FOR REFACTORING M13

In parallel to the genetic engineering project described above, students engaged in restructuring of the M13KO7 genome to make it more generally useful. Students could set their own design goals for this project, and, consequently, a wide range of ideas was generated. Several students chose to disentangle overlapping genomic elements, such as the overlapping p2 and p10 protein coding regions or the 3′ end of the p9 gene that is downstream of the 5′ end of the gene for p8. Other students pursued templates that could be useful to make unusual M13-based biomaterials. For example, several students chose to duplicate or alternatively regulate the gene for p8 to enable display of distinct forms of that protein on the phage coat. A third class of designs was intended to provide

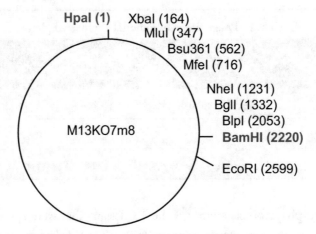

FIGURE 3.3: M13KO7m8 scaffold for refactoring. Terms in red type are natural sites, all others added by Felix Moser with silent mutations. Based on the design by Bryan Hernandez.

the phage with novel functions, such as a phage that could bridge two cells through a peptide fusion to the p9 protein, allowing interaction with the bacterial pilus. The final class of modifications included those that might enable better understanding of the phage life cycle, for example, the fusion of a fluorescent molecule to a phage protein to better visualize the localization and expression pattern of that protein during infection.

This wide range of design goals was pursued despite the limitations we imposed on the genomic "playing field." Only the first 2221 bases of the 8669-base M13KO7 genome were eligible for redesign. The reasons for this restriction were both technical and pedagogical. We believed the refactoring or reengineering of the genome from start to finish was too unwieldy a task to assign in the context of an undergraduate laboratory class. Even the successful T7 refactoring effort that inspired the M13 project did not include the entire genome of the T7 phage (Chan, Kosuri, & Endy, 2005). Importantly, the limited region of the M13 genome contains 7 of the phage's 11 protein-coding sequences, each with multiple opportunities for optimization. Thus, there seemed ample room for students to find unique projects, and the range of final proposals supports this assumption. Finally, unique restriction sites, namely, *Hpa*I at base position 1 and *Bam*HI at position 2221, flanked the start and end points for this region of the M13KO7. Unfortunately, there existed only two other unique restriction sites internal to the *Hpa*I and *Bam*HI sites, and thus before assigning this project to our undergraduate class, a scaffold was designed and built so small regions within the allowed 2221 bases could be manipulated.

The scaffold, called M13K07mut8, was made in the Endy laboratory to facilitate the students' refactoring work (Figure 3.3). We enlisted the help of a more senior biological engineering student, Bryan Hernandez, to identify eight additional restriction enzyme sites that might arise

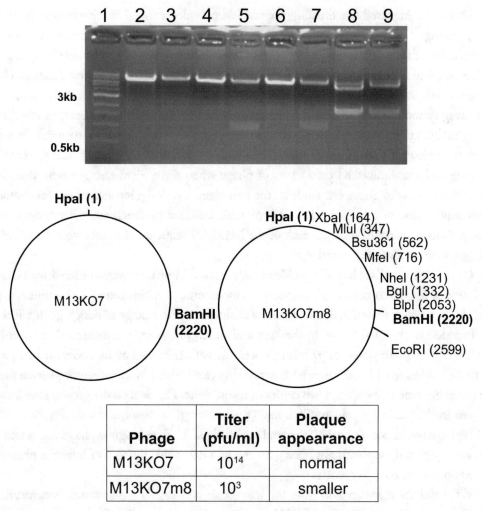

FIGURE 3.4: (Top) Scaffold DNA digested with HpaI+MfeI (lanes 2 and 3), Hpa1+NheI (lanes 4 and 5), Hpa1+Bg1I (lanes 6 and 7), HpaI+BamHI (lanes 8 and 9). (Bottom) Results of plaque assay with M13 phage of indicated genomes.

through silent changes within the M13KO7 genome. The mutations needed to add these eight new sites did not change the resulting protein sequences nor did they affect known phage "parts"-like promoters or ribosome binding sites. A technician in the Endy laboratory, Felix Moser, made the necessary sequence changes through a combination of direct synthesis and cloning as well as site-directed mutagenesis. The scaffold, with eight new unique restriction sites dotting the relevant portion of the genome, was successfully assembled. However, for unknown reasons, the phage appeared to grow orders of magnitude less well than the starting M13KO7 (Figure 3.4).

Despite its impaired functionality, the scaffold provided a useful starting canvas for the students' planning. The scaffold emphasized the value of modularity for engineering systems, both from the point of view of planning new designs and from the standpoint of trouble-shooting them later. The scaffold further emphasized the idea of biological parts and their standardization. Having considered both the benefits of parts standardization and the obstacles it presented to the operation of the living system, we did away with any requirement that the students adhere to a standard for parts assembly. A critic could reasonably charge that we gave up the ship too soon on this stated goal of synthetic biology. We took cues, however, from both the conservative approach taken for the T7.1 refactoring and the diminished production of phage when eight silent changes were introduced. Thus, in the interest of giving our students the best chance at designing something functional, we did not require that they bracket all their parts with standard BioBrick ends. For demonstration purposes, though, standard parts for each of the M13KO7 sequence elements were deposited into the Registry of Standard Biological Parts.

Of the dozens of student design ideas, only a handful of them were ordered for synthesis. The decisions for which designs to pursue were based on a combination of cost estimates, prior knowledge in existing scientific literature, and a desire to pursue a range of design goals within the class. Included in the final list of synthesized and tested phage were constructs that "unstuffed" genes for p9 from p8 or genes for p7 from p9, a phage with enhanced p9 functionality (e.g., with a YFP fusion), a phage with enhanced p3 functionality (e.g., with a biotin acceptor peptide fusion), and a phage genome that encoded two copies of the p8 gene. The dual p8 design was later found to have been in the published primary literature (Enshell-Seijffers, Smelyanski, & Gershoni, 2001), and this phage was tested along with the students' designs. Each redesigned phage was tested with a plaque assay and, disappointingly, none gave rise to a detectable number of infective phage with the exception of the dual p8 construct.

What did the students gain from the large-scale and largely unsuccessful reengineering of the M13 genome? The abstraction of DNA sequence elements into functional parts is a powerful framework for understanding because it requires that the students understand the DNA elements that encode any given function to "black box" them correctly. It further requires that the students understand what parts assemble into functional devices and what scientific knowledge they can bring to bear on devices as they design them. First-hand experience reengineering the M13 genome also offered a better appreciation of the challenge and success in the T7.1 refactoring project.

These outcomes were not merely a "felt sense," however. Instead, we took the opportunity to study in more depth what students learned from their M13 experiences and from the class as a whole. One aspect we were particularly interested in studying were the ways that students learned about M13 by communicating that learning in written and oral forms. As we describe in the next chapter, the effort to educate the next generation of biological engineers included giving students experiences in the discursive practices of professionals, namely, writing up their research in several forms. Writing as a tool of learning science and of communicating that learning has a long tradition (Bazerman, 1988), and studying this writing and its production provided us an opportunity to view the success of the M13 reengineering effort beyond the contribution to biological engineering itself.

REFERENCES

Bazerman, C. (1988). Shaping written knowledge: The genre and activity of the experimental article in science. Madison, WI: University of Wisconsin Press.

Browning, L. D. (1988). Interpreting the Challenger disaster: Communication under conditions of risk and liability. *Organization & Environment, 2.3–4*, pp. 211–227. doi:10.1177/108602668800200303

Chan, L. Y., Kosuri, S., & Endy, D. (2005). Refactoring bacteriophage T7. *Molecular Systems Biology, 1*, p. 2005.0018. doi:10.1038/msb4100025

Enshell-Seijffers, D., Smelyanski, L., & Gershoni, J. M. (2001). The rational design of a 'type 88' genetically stable peptide display vector in the filamentous bacteriophage fd. *Nucleic Acids Research, 29*(10), p. E50-0. doi:10.1093/nar/29.10.e50

Guthrie, R., & Shayo, C. (2005). The Columbia disaster: Culture, communication & change. *Journal of Cases on Information Technology, 7.3*, pp. 57–76.

Lima, M. (2007). Engineering education in the wake of Hurricane Katrina. *Journal of Biological Engineering, 1*, p. 6. doi:10.1186/1754-1611-1-6

Ph.D. peptide display cloning system (E8101), phage display, NEB. Retrieved March 19, 2009, from http://www.neb.com/nebecomm/products/productE8101.asp.

Talk:20.109(F07):Start-up genome engineering—OpenWetWare. Retrieved March 19, 2009, 2009, from http://openwetware.org/wiki/Talk:20.109(F07):Start-up_genome_engineering

van Wezenbeek, P. M., Hulsebos, T. J., & Schoenmakers, J. G. (1980). Nucleotide sequence of the filamentous bacteriophage M13 DNA genome: Comparison with phage fd. *Gene, 11*(1–2), pp. 129–148.

· · · ·

CHAPTER 4

Writing and Speaking as Biological Engineers

4.1 INTRODUCTION

As the previous chapter demonstrates, our experiences with teaching synthetic biology lead us to what might be categorized as a statement of the obvious: Students learn to be biological engineers by engaging in the actual practices of biological engineers. While key practices are the experimental work we have described, other essential activities include making sense of and communicating that experimental work, both in oral and in written forms. For students to learn synthetic biology, then, they need to learn *to be* biological engineers by engaging in the communication activities central to that pursuit, and this identity formation is a key goal in the learning outcomes we describe in this chapter.

This emphasis on learning science by communicating science is not a new phenomenon. Writing-across-the-curriculum or writing-in-the-disciplines programs have had a profound effect on American higher education in the last 25 years (Bazerman, et al., 2005). Nevertheless, science educators have struggled to enact this emphasis on communication in class and laboratory settings in which conveying technical or scientific content seems the top priority. Students have long written about their laboratory inquiry, but that writing has consisted of, in Hodson's words, "an almost unrelieved diet of worksheet-driven 'cookbook exercises' in which students slavishly follow teacher directions and only very occasionally engage in serious thought about what they are doing or why they are doing it" (1998, p. 93). As we noted in the previous chapter, "authenticity" has been a strong goal for the exploration of genome refactoring in class and laboratory settings at MIT, and the communication of that research in authentic forms is essential to overcome the limitations that Hodson describes.

Finally, the student writing and speaking that we describe in this chapter occur in the class of Laboratory Fundamentals of Biological Engineering, which is situated within a curricular context at MIT. We first offer some detail of that context both to situate our students' learning and to offer the factors that would need to be attended to when engaging in these activities in other settings.

4.2   COMMUNICATIONS-INTENSIVE CLASSES AT MIT

At MIT, all undergraduates are required to take four classes that are designated as "communications-intensive." Two of these classes meet a distribution requirement in the humanities, arts, and social sciences (CI-H), and the other two are classes within the major (CI-M). Thus, every department needs to offer its undergraduate majors two significant experiences in communication in that discipline—with, ideally, one building on the other (for more information, see http://web.mit.edu/commreq). Many of these CI-M classes have been mapped on to laboratory classes or ones in which students were generating data, writing up that research, and working in pairs and small groups, thus allowing for individual and small-group instruction as opposed to only whole-class lecture.

The curricular reform to bring about the current communications-intensive (CI) course requirement was the result of alumni feedback gathered in the mid-1990s. Although MIT alums felt that they had received top-notch technical educations, their lack of proficiency with writing and speaking were significant hurdles to professional success. In response to this feedback, in 2000 MIT faculty passed an Institute-wide initiative with the intention to integrate "substantial instruction and practice in writing and speaking into all four years and across all parts of MIT's undergraduate program" (Office of the Communication Requirement, 2008).

In terms of instructional support, MIT's Writing Across the Curriculum office within the Program in Writing and Humanistic Studies acts as a consulting agency to the individual departments. What this consultation looks like can vary from department to department: In some instances, stand-alone scientific communications courses complement large introductory laboratory classes. In others, students meet weekly to write up and present research that they have completed independently in previous semesters, and, as is the case in Laboratory Fundamental of Biological Engineering, writing and speaking staff work within the existing class structure, offering students feedback on their communications work, grading writing and speaking tasks, and being part of the larger instructional team.

Overall, CI courses at MIT emphasize communication in the learning of technical content. Unlike some universities where such courses might be taught as stand-alone entities or with minimal input from the Writing Program, MIT's CI courses are taught collaboratively with technical and writing faculty. Each department at MIT develops its own CI courses to reflect the disciplinary needs of its students, some of whom enter industry and some of whom continue to graduate or professional school. This integrated approach develops students' writing and speaking skills in the practice of doing science and engineering.

4.3   LEARNING AND THE DEVELOPMENT OF PROFESSIONAL IDENTITY

As we noted, the writing and speaking that students do in CI classes are ideally based on the authentic practices of professionals in their fields. This approach is an apprenticeship model, in which students are developing identities as scientific professionals. The relationship between identity and learning has been seen as a key component to attend to if students' learning outcomes are to be successful (Gee, 2000–2001; Lave, 1996; Lave & Wenger, 1991). In other words, as students learn the technical content of science or engineering, that learning takes place within a host of social contexts—whether the laboratory, the classroom, the university—is enabled by social interaction with peers, instructors, other mentors, and is also interwoven with students' sense of their career and professional goals or with who they are as students and neophyte scientists (for an example of this approach in action, see Hanauer et al., 2006).

The use of writing and speaking tasks to teach genome refactoring offers rich opportunities to tap into these key elements. On the most obvious level, asking students to communicate their learning offers opportunity to assess what they have learned. In addition, communicating their learning puts students in the role of scientific professionals, having not merely to communicate but to persuade readers/listeners why their ideas or intepretations are valid, scientifically sound, and useful. Engaging in these essential communicative practices of professionals allows students to "try on" the identities of biological engineers, supported with mentoring and instruction. Thus, students are forming "discursive identities" (Brown, Reveles, & Kelly, 2005)—or the scientific identities they create through communicating as a scientific professional—and this process results in meaningful

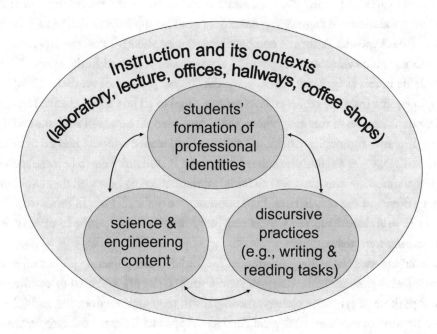

FIGURE 4.1: A model for learning to communicate in biological engineering.

learning. It is why students identify writing as vital to their educational experiences (Light, 2001) and also identify as more meaningful the writing and speaking they do in classes in their majors than writing in general education curriculum (Beaufort, 2007; Carroll, 2002; Sternglass, 1997). We summarize this model of learning in Figure 4.1, showing the interrelationship between scientific content, student writing, and discursive identity, all of which are mediated by instruction and take place in a variety of instructional contexts (e.g., the laboratory, lecture, faculty offices, hallways, coffee shops).

4.4   WRITING AND SPEAKING ABOUT SYNTHETIC BIOLOGY

In our context for teaching synthetic biology, we have asked students to communicate their learning through oral presentations and writing assignments, as a means for students to learn technical content and also as a way for students to build identities as biological engineers and to be immersed in the varied communicative practices of professionals. One example is a homework assignment we have used in which students offer a one-paragraph response to the statement: "Synthetic biology is about engineering while genetic engineering is about biology." By thinking about and expressing the similarities and differences between these fields, students need to articulate reason for synthetic biology itself to exist and begin to imagine their roles within that field.

The more formal and graded writing tasks that students do based on their attempts to reengineer M13 have varied. In spring 2007, we asked students "to write a thoughtful, researched essay exploring how a foundational engineering concept . . . can be applied as a design tool for biological engineering" (see Appendix A for the complete assignment sheet). Thus, the laboratory work on M13 acted to generate evidence to support the argument as presented in the essay. The necessary step for students in this assignment was to grasp the "bigger picture" of synthetic biology and how refactoring a genome from the sequence level up compared to ad hoc approaches to reengineering a biological system. Success in this assignment, then, depended on how well students could embrace these larger ideas and communicate them to a fairly general reader, one who needed to be convinced that synthetic biology was a viable research methodology. If students were to become spokespersons for biological engineering, this essay was an initial attempt at seeing how well they could muster the arguments to carve out the resources and attention necessary for success. In other words, students needed to write in an identity as researchers convinced that the field and the laboratory techniques they were learning were viable.

In a later semester, the writing assignment for the M13 reengineering project offered two additional angles. This time students were given two writing tasks: (1) a rebuttal to an editorial on the viability and purpose of synthetic biology in comparison to genetic engineering and (2) a business plan for the Registry of Standard Biological Parts (see Appendix B for the full assignments). As was true in the previous semester, students would use their research results on reengineering the M13 genome as particular forms of evidence for making these larger arguments, both of which called for addressing relatively general readers.

In terms of speaking tasks, students usually have two formal graded oral presentations. The first is a presentation of a research article, modeled on the "journal-club" activity common to scientific research groups. Students' presentations are videotaped and then in one-to-one conferences, students meet with a communications instructor to review their performance and receive feedback. The second oral presentation is done in pairs and is a proposal for a significant project applying the techniques of synthetic biology to a real-world problem or question. Some previous examples of these projects include a detoxifying agent based on the refactored M13, the reengineering of phage for oil-spill remediation, and the engineering of *Saccharomyces cerevisiae* for processing biomass into biofuels.

4.5   UNDERSTANDING STUDENTS' EXPERIENCES WITH COMMUNICATING SYNTHETIC BIOLOGY

The question of what it meant for students to learn to write in this class interested us in a variety of ways, whether as a means of improving instruction and outcomes or as a way of exploring the relationship between writing, speaking, and developing professional identity. Through whole-class surveys, periodic interviews with instructional staff and eight targeted students, and analysis of those eight students' written work from first to final draft, we explored students' experiences with conducting novel scientific research and how those experiences affected their satisfaction with the course as a whole (for additional exploration of this question, see Lerner, 2009, chap. 8; Craig, Poe, & Lerner, 2008).

Responses surprised us: Some students acknowledged that it could feel somewhat artificial to be writing a paper on research that had uneven results. However, students also seemed to have a grounded perspective on the ways the M13 work fit into the larger scheme of genome engineering as well as in the larger scheme of their futures.

One issue is the time available for the laboratory work in the class, particularly the necessary acceleration of the process from initial experiments to final paper. Some students did make note of this phenomenon, with one commenting that

> in the real world you get a little longer of a timeline for coming up with an idea to refactor a genome or writing up a lab report or coming up with a research proposal. So I mean, I feel like I'd like real life better, but this was kind of real life or your entire research career compacted into one semester.

Another issue was the relative success of the lab work and the need to obtain results from other lab partners and to forge ahead. As one student noted,

> The problem with our M13 work is that . . . there was some contamination, and there was a little mishap with the techniques that we were doing, . . . we had our own data to analyze, and we had another team's data to analyze as well that we borrowed. So throughout the whole lab report, the write up was a little difficult, and I guess now that we realized that, because we kept having to switch back and forth and analyze this and see what would happen with that, or take their results and kind of backtrack a little bit and see where they came from.

Nevertheless, many students were able to put into perspective the ups and downs of the M13 work, fully realizing that they were in a learning situation in which process was the focus, not necessarily the products of experimental work. For example, one student noted that the effect of his uncertain M13 results "wasn't nearly as much as I expected. A lot of it was mainly extrapolating on our thoughts on how this applied to engineering and what we did in lab was basically an introduction to the idea." In this introduction, students were learning the challenges to be overcome through subsequent experiences as well as their potential roles in those futures.

In terms of overall satisfaction with the class, students consistently described positive outcomes, often framing a semester's worth of experience in terms of their future goals. As one student described, "It was a fantastic experience. The amount [of work] and certain papers [were] daunting, but overall the experience was a great one. The class was taught well and the challenges pushed me in a constructive manner."

Another student's response reflects her growing sense of identity *as* a biological engineer and the way the class prepared her for a role in this relatively new field:

> I had never realized how much impact I had being one of the first few students to major in biological engineering and graduate from this program and how this research is incredibly new. . . . And that's really exciting.

Students' view of the class and its communciation tasks as professional preparation was a strong theme in these responses. One student was able to take a broad view of the class and its role in her undergraduate and professional career:

> Because we did so much in this class, I don't think that [the class] was intended to make us experts on writing papers and giving presentations. It felt more like wetting our feet before diving in. But that's really what the [biological engineering] undergrad major is all about. . . .

One of my professors said that the intent of the major is to lower the barrier for us if we encounter these topics later in our careers.

Although all students inteviewed expressed a great deal of satifsfaction with their learning experiences, at least one (who happened to have earned the top grade in the class) did react to the sheer volume and pacing, noting that

> I definitely appreciate the fact that we had such a broad overview of everything that biological engineering . . . involves, like from, you know, aspects of writing scientific articles to the aspect of like trying to make a profit to the aspect of like what like, the material transfer agreement that we have to look at, and like what did that involve. . . . This class has covered so much more than, you know, any class that I've known and has been one of the most useful, I think, that I've taken at MIT, in terms of broader applications further on, but there's a point where it's, like, too much, where it's just, like, we haven't been taught this.

The lesson we take from this sort of response is not necessarily the need to ratchet back our expectations for student learning but instead to realize that learning needs to be supported as fully as possible, that explicit instruction is key, and that teaching staff are decidely in mentoring roles, presenting identities for students to model. These might be daunting challenges, but they are key ones to attend to if students are to contribute to biological engineering and to advance the work of the field.

4.6   CONCLUSION

Our experiences with designing curricula and teaching genome reengineering have offered us opportunities to discover and rediscover the many ways that students learn to be scientists. Our commitment to having students write and speak about their science is strengthened by these outcomes, for it is in the relationship between curriculum, instruction, and student activities that real learning takes place. Attention to this relationship occupies a great deal of time and attention, as well as institutional resources, but the goal of creating the next generation of biological engineers certainly demands that effort.

REFERENCES

Bazerman, C. (1988). Shaping written knowledge: The genre and activity of the experimental article in science. Madison, WI: University of Wisconsin Press.

Bazerman, C., Little, J., Bethel, L., Chavkin, T., Fouquette, D., & Garufis, J. (2005). *Reference guide to writing across the curriculum*. W. Lafayette, IN: Parlor Press.

Beaufort, A. (2007). *College writing and beyond: A new framework for university writing instruction.* Logan, UT: Utah State UP.

Brown, B. A., Reveles, J. M., & Kelly, G. J. (2005). Scientific literacy and discursive identity: A theoretical framework for understanding science learning. *Science Education, 89*, pp. 779–802. doi:10.1002/sce.20069

Carroll, L. A. (2002). *Rehearsing new roles: How college students develop as writers.* Carbondale: Southern Illinois UP.

Chan, L. Y., Kosuri, S., & Endy, D. (2005). Refactoring bacteriophage T7. *Molecular Systems Biology, 1*, p. 2005.0018. doi:10.1038/msb4100025

Craig, J. L., Poe, M., & Lerner, N. (2008). Innovation across the curriculum: Three case studies in teaching science and engineering communication. *IEEE Transactions on Professional Communication, 51.3*, pp. 1–22.

Gee, J. P. (2000–2001). Identity as an analytic lens for research in education. *Review of Educational Research, 25*, pp. 99–125. doi:10.2307/1167322

Hanauer, D. I., Jacobs-Sera, D., Pedulla, M. L., Crewsawn, S. G., Hendrix, R. W., & Hatfull, G. F. (2006). Teaching scientific inquiry. *Science, 314*, pp. 1880–1881. doi:10.1126/science.1136796

Hodson, D. (1998). Is this really what scientists do? Seeking a more authentic science and beyond the school laboratory. In J. Wellington (Ed.), *Practical work in school science: Which way now?* (pp. 93–108). London: Routledge.

Lave, J. (1996). Teaching, as learning, in practice. *Mind, Culture, and Activity*, 3.3, pp. 149–164. doi:10.1207/s15327884mca0303_2

Lave, J., & Wenger, E. (1991). *Situated learning: Legitimate peripheral participation.* Cambridge, UK: Cambridge UP.

Lerner, N. (2009). *The idea of a writing laboratory.* Urbana: Southern Illinois UP.

Light, R. (2001). *Making the most of college: Students speak their minds.* Cambridge, MA: Harvard UP.

Office of the Communication Requirement (2008). *About the Requirement.* MIT Undergraduate Communication requirement. Retrieved February 1, 2008, from http://web.mit.edu/commreq/background.html.

Sternglass, M. S. (1997). *Time to know them: A longitudinal study of writing and learning at the college level.* Mahwah, NJ: Erlbaum.

. . . .

CHAPTER 5

Summary and Future Directions

5.1 SUMMARY

Although the bacteriophages T7 and M13 have considerable differences in their life cycles and their genomic organization, both have been subject to genome refactoring and reengineering efforts. In the case of T7, the "leftmost" 11.5 kilobases of the linear genome were refactored to disentangle overlapping functional elements (Chan, Kosuri, & Endy, 2005). The resulting genome had a modular architecture, with each DNA sequence associated with one and only one function, and this effort gave rise to functional phage, albeit less robust (Figure 2.4) and, evolutionarily speaking, unfavored (unpublished)—a point we will return to below. In the case of M13, patches in the "first" 2.2 kilobases of the circular genome were refactored. A variety of design goals were intended, for example, to duplicate genes encoding phage coat proteins, to independently regulate open reading frames that are normally synched, and to tag particularly useful elements of the phage coat. With rare exception, M13 did not tolerate such manipulations, and most of the resulting genomes were unable to productively infect a bacterial host. Indeed, we observed that eight silent mutations decreased the number of productive phage infections by greater than 10 orders of magnitude (Figure 3.4). Thus, although T7.1 is widely considered a bold prototype, successful enough to bootstrap and improve, the frustrations from the M13.1 refactoring effort cast a sobering light on the challenge of genome redesigns, reminding us that nature is a harsh critic.

Given the availability of a refactored T7 genome, what engineering directions have opened as a result? Recall, the refactoring was motivated in large part by a desire to generate a more "model-able" organism for study. Computer models of the natural genome failed to accurately predict the consequence of gene reorganization, specifically the consequence of delaying the entry and expression of a key gene (the T7 RNA polymerase) until late in the infection cycle. It is reasonable to think that a more modular T7 genome like the refactored one could be deployed to address this discrepancy. In addition, the uncharacterized open reading frames from T7 that were identified by sequence analysis but that have no known functions could be more easily and more precisely removed from a refactored genome. In this way, all sequences could be characterized as meaningful (or not) for T7 biology.

The features of M13 that made it a good candidate for refactoring also made it an ideal template for teaching a new cadre of biological engineers. Given an opportunity to build a genetic program that meets explicit design goals, our students had hands-on experiences with the core concepts of genome reengineering and engaged in the practices of next-generation biological engineers, including the many writing and speaking tasks. As we noted in the previous chapter, on the whole, students found these experiences to make positive contributions to their futures.

5.2   FUTURE DIRECTIONS

Engineering disciplines can grow from advances in particular areas of science. For instance, electrical engineering grew from discoveries in physics. It is likewise possible that biological engineering can grow from a molecular and system-level understanding of the life sciences. The development is far from certain, however. Unlike physics, biology does not rigidly adhere to many known rules. It is possible that living cells are simply too varied and too complex to be standardized. Context dependencies are observed as biological parts are moved from one organism to another. Codon usage is known to differ in different cell types. Stress responses and tolerances differ. Even defined perturbations to a given type of cell can give rise to distinct outcomes because a cell's physiology can vary widely, and these variations can affect the reactions it undergoes. Engineers, when faced with complex problems, reduce the complexity by setting boundaries within which they work. For instance, bridges have load limits, and electronics have defined operational temperatures. Could similar operational parameters be defined for biological parts? To some extent, this has been successfully accomplished already. For example, bacterial shuttle vectors are categorized according to the copy number they confer and the compatability of their replication origins. To similarly characterize a wider range of biological parts, a large-scale effort would be needed, made exponentially more difficult if both individual and combinations of parts require characterization.

Even a collection of well-characterized, standard biological parts will carry a variable not known in other science and engineering fields, namely, that of evolution. Transistors remain transistors over every imaginable time scale and can reliably work in combination to support integrated circuits. By comparison, genetic control elements will accumulate random changes at a measurable rate, with consequences that are challenging to predict for the control element itself and for the system it contributes to. T7.1 provides a striking case in point because the refactored genome evolves to eliminate or reduce the human-designed elements (unpublished data). No other engineering field must contend with evolutionary pressures, and perhaps biological engineering would be best served by trying, whenever possible, to harness rather than limit its inevitable intervention. Indeed, directed evolution has been used to convert nonfunctional synthetic genetic circuits to functional ones, optimizing the components of a system in ways that were not predicted in advance (Yokobayashi, Weiss, & Arnold, 2002). Thus, synthetic biology may be useful to fuel both scientifically important discoveries and technologically driven applications of this knowledge.

Finally, as we have described here, genome refactoring offers a context for teaching that is in keeping with the goals and missions of many educational institutions. By offering an authentic context for scientific inquiry, the M13 genome reengineering project inspired our students to wrestle with uncertainty, to make decisions based on incomplete knowledge, and to interpret and communicate their findings in the context of what they expected, what others in the class discovered, and what information was published in the primary literature. In this way, their experience was much more akin to that of a practicing biological engineer than to that of a student in an introductory laboratory subject. What is particularly gratifying, to them as well as to us, is the observation that their data, albeit imperfect and often disappointing, have advanced our knowledge of genomic engineering, as evidenced by this lecture. We are excited to imagine many similar projects with such dual benefit.

REFERENCES

Chan, L. Y., Kosuri, S., & Endy, D. (2005). Refactoring bacteriophage T7. *Molecular Systems Biology, 1*, p. 2005.0018. doi:10.1038/msb4100025

Yokobayashi, Y., Weiss, R., & Arnold, F. H. (2002). Directed evolution of a genetic circuit. *Proceedings of the National Academy of Sciences of the United States of America, 99*(26), pp. 16587–16591. doi:10.1073/pnas.252535999

· · · ·

APPENDIX A

Genome Engineering Essay Assignment (Spring 2007)

NOTE: Major writing assignments associated with the M13 Genome Engineering experiments are fully detailed here. During other experimental modules associated with the class, students write about their laboratory work in laboratory reports and research articles, as well as by completing homework assignments, keeping laboratory notebooks, and engaging in some of the more traditional forms of school science writing. Retrieved from http://openwetware.org/wiki/20.109(S07):_Genome_engineering_essay

You are asked to write a thoughtful, researched essay exploring how a foundational engineering concept (e.g., abstraction, modularity, insulation, standardization, decoupling) can be applied as a design tool for biological engineering. Your lab work with M13 will provide the context for your argument.

ABSTRACT

This has been written for you to clarify the assignment. You can include this abstract as your own.

To engineer novel biological systems, we need to change the genetic code of existing biological materials, not by making a few changes as current methods allow us to do but rather by making lots and lots of changes in a fast, cheap and reliable way. Just as "plug-ins" provide new or improved functions to existing computer programs, the current tools of molecular biology allow for piecemeal modification to genetic programs, adding functionality but often complexity and clumsiness as well. In this essay I will describe two approaches to biological programming, *ad hoc* adjustment and complete refactoring, as applied to the simple genome of the bacteriophage M13. With both approaches, I will show how the application of a foundational engineering concept, namely (abstraction, insulation, standardization, decoupling, modularity . . . choose one), enables more reliable and elegant genetic programming and can give rise to a platform with more flexibility and fewer restrictions.

INTRODUCTION

From your introduction, your readers expect to find out why your topic is important and why they should be interested in it. To do that, you need to describe the larger context for the work, the ways it's important, and the specific areas your paper will address. There's no need to hide your main point or approach. At the end of the introduction, the reader should want to learn how the foundational idea you've chosen (abstraction, modularity, insulation, standardization or decoupling) serves a useful purpose and affords great opportunity if incorporated among "best practices" for biological engineers, expecting M13 to be the test case they'll follow.

Launch this section using one of the following quotes, or a personal favorite.

- Today, most software exists, not to solve a problem, but to interface with other software. (I. O. Angell)
- Programming languages should be designed not by piling feature on top of feature, but by removing the weaknesses and restrictions that make additional features appear necessary. (Anonymous, Revised Report on the Algorithmic Language Scheme)
- Programs for sale: Fast, Reliable, Cheap: choose two. (Anonymous)
- Think (design) globally; act (code) locally. (Anonymous)
- Think twice, code once. (Anonymous)
- Weeks of programming can save you hours of planning. (Anonymous)
- Any fool can write code that a computer can understand. Good programmers write code that humans can understand. (M. Fowler, "Refactoring: Improving the Design of Existing Code")
- A program like Microsoft's Windows 98 is tens of millions of lines of code. Nobody can keep that much complexity in their head or hope to manage it effectively. So you need an architecture that says to everyone, "Here's how this thing works, and to do your part, you need to understand only these five things, and don't you dare touch anything else." (C. Ferguson "High Stakes, No Prisoners; Times Business Press")

Explicitly describe what problem or issue the quote you've chosen highlights and how the point applies to genetic programming as well.

Next . . . well, it's really up to you. You could

- Allow one of the more familiar software disasters to illustrate comparable design problems that can be encountered when making biological materials
- Describe some (but not all) current practices in genetic programming and their limitations
- Introduce M13 as the example you've chosen to hack and debug

Some ideas you may want to introduce are listed but this list is neither exhaustive nor mandatory.

- Complexity
- Managing complexity
- Simplicity
- Refactoring
- Features of good/bad computer programs
- Features of good/bad genetic programs
- Methods for testing and debugging
- Engineering
- Science
- Understanding
- Standardization
- Dynamics
- Decoupling
- Abstraction
- Evolution
- Usefulness
- Discovery

BODY: PARTS 1–3

In these sections you will build off of your introduction to present M13 as an example of the issues you've highlighted. Your readers expect to learn something from what you present; thus, you'll need to supply ample description as well as an analysis of your lab results. Remember your goal is to make a persuasive argument for the concept of abstraction (or modularity, insulation, standardization, decoupling . . .) with evidence from your laboratory experience.

Part 1. How it's built: M13 as a test case

At the conclusion of this section, the reader should have a good understanding of

- The prevalence and diversity of bacteriophage
- The M13 life cycle (include a figure if you like)
- The size and organization of the genome
- The proteins encoded by the genome structure (include a figure or table if you like)
- Any natural variations to the genome

End this part by highlighting how "engineerable" the natural example seems and how (abstraction, modularity, insulation, standardization or decoupling) is the key to reliably and predictably accomplishing this.

Part 2. Build to learn: M13 and piecemeal fixes

At the conclusion of this section, the reader should have a good understanding of

- The application of M13 for phage display, cite at least one successful application of this technique
- The limitations/variations of phage display
- The modification to the genome that you performed in lab and what useful purpose it could serve
- Your plaque assay and Western data, be it positive or negative (include a figure and table)

End this part by commenting on how fast, cheap, and reliable this approach proved to be. On the scale of other engineering feats, how ambitious was it? How much expertise was required? How can you imagine making it an easier and more robust engineering task?

Part 3. Learn to build: refactored M13

At the conclusion of this section, the reader should have a good understanding of

- What refactoring is
- What the rough draft of refactored M13 tried to do
- Which gene (gII, gIX, gVIII, gIII, or gXI) you carefully refactored and how you approached then solved the specifics of that problem
- Your plaque assay and Western data (when it is available), be it positive or negative (include a figure and table)

End this part by commenting on how refactoring compares to ad hoc tweaking and how much or how little promise it holds for building fast, cheap, and reliable biological systems.

CONCLUSIONS OR SUMMARY

In this section, your readers expect you to tie up the concepts you raised in your introduction with the specific examples you've described in terms of M13. Most important, you need to supply some "future thinking" about the implications of what you've presented, whether for future experimental work or the larger field.

APPENDIX B

Genome Engineering Assessment (Fall 2007)

PART 1: REBUTTAL TO EDITORIAL

This will be written as a homework assignment, exchanged with your lab partner for peer review, and then submitted to the teaching faculty as part of your portfolio. This portion of the assignment accounts for up to 15% of your grade.

Choose **one** of the following **two** essays to write a thoughtful response to their "fighting words." Rebut the quoted statements by first explaining what the quote refers to, explaining why the author or quoted individual might have said it, and then provide at least five counterpoints or examples to support the opposite point of view. Draw your arguments from your experiments with M13 whenever possible. Print out two copies of this portion of the assignment. Next time you and your lab partner will exchange responses and provide feedback to each other on the writing and ideas within.

Essay choice 1: Choose ONE of the following quotes to address. Both come from Andrew Pollack in the *New York Times*, Tuesday, Jan 17, 2006, Custom-Made Microbes, at Your Service, which quotes Professor Arnold of Caltech as saying:

- "(Synthetic Biology) has a catchy new name, but anybody over 40 will recognize it as good old genetic engineering applied to more complex problems."

and

- "There is no such thing as a standard component, because even a standard component works differently depending on the environment. The expectation that you can type in a sequence and can predict what a circuit will do is far from reality and always will be."

Essay choice 2: Editorial: Meaning of Life in *Nature* (2007) 447: 1031–1032:

- "it would be a service . . . to dismiss the idea that life is a precise scientific concept"

PART 2: MINI-BUSINESS PLAN FOR THE REGISTRY OF STANDARD BIOLOGICAL PARTS

Put yourself 5 years in the future and imagine that the Registry is floundering. Though the number of useful parts has grown through the hard work and dedication of its volunteer workforce in the iGEM program, there is a notable lack of standards:

- Around the parts themselves (some work always, some in rare conditions, some not at all)
- Around the assembly process (alternative biobricks and registries have gained popularity)
- And around documentation for the parts (some have great spec sheets and some have nothing).

Decide that you will direct the Registry into a manufacturing, service, high tech, or retail business and then devise a plan to grow and stabilize that business. This portion of the assignment accounts for up to 25% of your grade.

In no more than three pages provide a business plan that includes:

1. An executive summary in 250 words or fewer, explain:

- What is your product
- Who are your customers
- What the future holds for the registry in particular and synthetic biology more generally
- What you see as the key to success

This summary should sound enthusiastic, professional, and be more readable than most "mission statements." Also, consider writing this section after you've written the rest of the plan.

2. Summary of the current Registry

- Describe what the Registry is, including products, services, customers, ownership, history, location, facilities.
- Include strengths and core competencies of the Registry.
- Segue into the next section by mentioning the significant challenges faced in the near and long term.
- This section should be no longer than two paragraphs.

3.   Market analysis

Dedicate one paragraph to a description of the market. You might consider including information like:

- Who makes up your market?
- What is its size now? How fast is it growing? How do you know?
- What percentage of the market do you expect the Registry to have now and 5 years from now?
- How could changes in technology, government, and the economy affect your business?

4.   Business plan

Specify your strategy for continued growth of the Registry. The emphasis of this section will differ depending on the kind of business model you have chosen (retail, manufacturing, service or high tech).

Here are some questions you might consider as you formulate your business plan:

- How will you promote the use of the Registry?
- How will you advertise?
- How will you price your product/services?
- Where will you locate the Registry (or BioBrick franchises) and how you will distribute parts/services?
- How you will keep the Registry competitive?
- How/if you will protect intellectual property while also promoting sharing and community?
- Does your plan emphasize increased production, diversification, or eventual sale of franchises?
- How long will your strategy take to be partially or fully realized?
- Are there start-up costs associated with your business model? How much and where will the capital come from?
- Will your registry require insurance coverage or litigation insurance?
- Are there trademarks, copyrights, or patents (pending, existing, or purchased) considerations?
- How many and what kind (skilled, unskilled, and professional) of employees to you anticipate?
- Where will you recruit employees?
- Will top notch employees advance? To what?

- How will you training employees?
- What kind of inventory will you keep: raw materials, supplies, finished goods?
- Will there be seasonal fluctuations to demand for parts?
- Will you need lead-time for ordering?
- Do you expect shortages or delivery problems?
- Are supply costs steady? Reliable?
- Will you sell parts on credit?
- How will you set prices?
- What kind of guarantees and privacy protects will you offer?

This section has no defined length or format but should end on an enthusiastic note that might lead some venture capital firm or a funding agency to stay interested.

APPENDIX C

Nuts and Bolts of Molecular Biology

SECTION 1: VOCABULARY

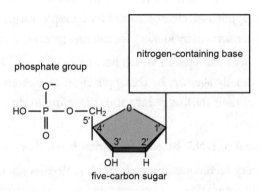

This picture is from: http://www.biology.iupui.edu/biocourses/N100H/images/3nucleotide.gif

5′: Carbon on sugar of nucleotide that is covalently linked to a phosphate group

3′: Carbon on sugar of nucleotide that is covalently linked to a hydroxyl group. Addition of new nucleotides to growing DNA or RNA chains occurs at hydroxyl so elongation is said to proceed in 5′ to 3′ direction.

Amber mutation: change in a DNA codon to TAG, leading to UAG in RNA and a premature stop codon. This mutation can be suppressed in some bacterial strains with altered tRNAs. Such strains are called amber suppressor strains.

Central dogma: DNA is the template for RNA is the template for protein.

Codon usage: preference for one codon over another when both code for a single amino acid. Because codon usage varies between some species, the transfer of a gene from one species to another can be complicated if the gene is not recoded according to the codon preference of the new host.

Direct repeats: two sequences of DNA that are identical and aligned head to tail, with or without intervening DNA that is not part of the repeating units. When such repeated sequences are on the order of two dozen base pairs, then homologous recombination can occur, deleting any DNA sequence in between the repeats.

Gene: a segment of DNA that specifies a protein through an mRNA intermediate, a tRNA, or an rRNA. The word "gene" can sometimes refer to the entire expression unit, which includes promoters, regulatory sequences, etc, and can sometimes refer to the open reading frame only.

Genetic screen: an experiment in which a cell or organism with particular characteristics is identified from a larger population. To increase the odds of finding such a cell or organism, the population is often mutagenized in advanced of screening.

Genetic selection: an experiment like a genetic screen except that only cells or organisms with a desired characteristic can survive the applied selective pressure. One example of a genetic selection is the identification of antibiotic resistant microbes by looking for colonies growing on media with the antibiotic.

Gel electrophoresis: a laboratory technique in which negatively charged DNA molecules are drawn through a sieving-matrix, usually agarose, by the application of a current. The distance migrated by the DNA molecules from their starting point varies according to their length, with the smallest molecules migrating furthest.

Genome: the total complement of DNA-based information in a cell or virus.

Northern analysis: a laboratory technique in which RNA molecules are drawn through a sieving-matrix, usually agarose, by the application of a current and then the separated RNAs are transferred from the matrix to a membrane, often nitrocellulose. RNAs that are bound to the membrane are then probed with radioactively-labeled DNA-probe to detect the migration and relative amount of a particular RNA of interest.

Oligonucleotide: a short sequence of single-stranded DNA (usually on the order of 20-60 nucleotides). These pieces of DNA can be synthesized to a specified sequence and are used in PCR, and is then called a "primer."

ORF: "open reading frame," the DNA pattern of triplet sequences that encode a protein.

Origin of replication: the necessary and sufficient DNA sequence to initiate replication.

Promoter: sequence of DNA to which RNA polymerase binds for initiation of transcription.

Restriction enzyme: an enzyme that recognizes and cleaves a specific DNA sequence.

Restriction site: the sequence of DNA recognized and then cleaved by a microbial endonuclease. The sequence itself is almost always palindromic, meaning it is the same read from 5′ to 3′ on either strand.

Replication: the reaction that generates a new copy of DNA from an existing copy. This reaction is catalyzed by DNA polymerase.

RBS: "ribosome binding site," the sequence of RNA to which ribosome binds for initiation of translation.

Restriction Enzyme	Restriction Site (5'→3' With The Point Of Cleavage Indicated By An ^)
*Bam*HI	G^GATCC
*Eag*I	C^GGCCG
*Eco*RI	G^AATTC
*Hpa*I	GTT^AAC
*Kpn*I	GGTAC^C

Site-directed mutagenesis: a method to intentionally change a DNA sequence, usually converting one codon to another.

Start codon: the codon at which translation begins, most often AUG.

Stop codon: the codon at which translation ends, most often UAA, UAG, or UAG. These are sometimes called "nonsense codons."

Transcription: the reaction that converts of DNA-templated information to RNA. This reaction is catalyzed by one of several RNA polymerases.

Transcriptional terminator: a sequence of DNA that signals the RNA polymerase to cease the synthesis of RNA. Terminator sequences are often inverted repeats in the DNA that fold into stem-loop structures, leading the RNA polymerase to pause and leave the DNA it is transcribing.

Translation: the reaction that converts RNA-templated information to protein. This reaction is catalyzed by ribosomes.

Western analysis: a laboratory technique in which protein molecules are drawn through a sieving-matrix, usually polyacrylamide, by the application of a current and then the separated proteins are transferred from the matrix to a membrane, often nitrocellulose. Proteins bound to the membrane are then probed with a solution of antibody to detect the migration and relative amount of a particular protein of interest.

SECTION 2: MOLECULAR BIOLOGY
Generally useful sites:
http://www.ncbi.nlm.nih.gov/About/primer/genetics_molecular.html
http://www.teachersdomain.org/collection/k12/sci.life.gen.mechdna/
http://www.dnalc.org/home_alternate.html

DNA Sequencing

http://openwetware.org/wiki/20.109(F07):_M13.1

At the heart of sequencing reactions is chemistry worked out by Fred Sanger in the 1970s that uses dideoxynucleotides. These chain-terminating bases can be added to a growing chain of DNA but cannot be further extended. Performing four reactions, each with a different chain-terminating base, generates fragments of different lengths ending at G, A, T, or C. The fragments, once separated by size, reflect the DNA's sequence. In the "old days," radioactive material was incorporated into the elongating DNA fragments so they could be visualized on X-ray film. More recently, fluorescent dyes, one color linked to each dideoxy-base, have been used instead. The four colored fragments can be passed through capillaries to a computer that can read the output and trace the color intensities detected. The invention of automated sequencing machines has made sequence determination a fast and inexpensive endeavor.

DNA Synthesis

http://www.biobuilder.org/dnasynthesis.html

Chemical Synthesis of Oligonucleotides

http://www.idtdna.com/TechVault/TechVault.aspxor

The chemistry of DNA synthesis involves protection and deprotection of a growing DNA chain with specialized bases called deoxynucleoside phosphoramidites. Only one base can link to a DNA chain until the end gets deprotected, which then allows the next base to be added to the growing sequence. A "capping" step follows the addition step so that any deprotected base that does not properly connect to the next base in the chain cannot be further extended. The shorter pieces of DNA can be taken out of the mix at the very end. The construction process is more than 99% efficient, allowing for routine synthesis of oligonucleotides that are 20–60 bases. Synthesis companies verify the final DNA by sequencing it before they send it to the end user.

Ligation and Transformation

http://openwetware.org/wiki/20.109(F07):_DNA_ligation_and_bacterial_transformation

The goal of ligations is to combine two pieces of DNA that have been cut with restriction enzymes so as to leave compatible ends. During ligation reactions, hydrogen bonds will form between the overhangs on the fragments, and then the ligase will repair the phosphate backbone, creating a stable circular plasmid.

Transformation is the term given to the process of adding DNA, that is, ligation reactions, into bacteria. During "transformation," a single plasmid from the ligation mixture enters a single bacterium and, once inside, replicates and expresses the genes it encodes. In the case of M13KO7, one of the genes on the genome leads to kanamycin resistance. Thus, a transformed bacterium will grow on agar medium containing kanamycin. Untransformed cells will die before they can form a colony on the agar surface.

Most bacteria do not usually exist in a "transformation ready" state, but the bacteria can be made permeable to the plasmid DNA, and cells that are capable of transformation are referred to as "competent." Competent cells are extremely fragile and should be handled gently, specifically kept cold and not vortexed. The transformation procedure is efficient enough for most laboratory purposes, with efficiencies as high as 10^9 transformed cells per microgram of DNA, but in fact, even with high efficiency, cells only 1 DNA molecule in about 10,000 is successfully transformed.

PCR

http://openwetware.org/wiki/20.109(F08):_Mod_1_Day_1_DNA_engineering_using_PCR
http://www.ncbi.nlm.nih.gov/About/primer/genetics_molecular.html

PCR is the acronym for "polymerase chain reaction." The goal of any PCR is to generate many copies of DNA from a few. This technique is useful for amplifying DNA of known or unknown sequence. The reactions require only four components: DNA to be amplified (called the "target" or "template"), oligonucleotide primers to bind sequences flanking the target, dNTPs to polymerize into new DNA chains, and a heat-stable polymerase in a buffered solution to carry out the synthesis reaction over and over and over. PCR is a three-step process (denature, anneal, extend), and these steps are repeated 20 or more times. In the denaturing step, the reactions are heated to 95°C, which melts the DNA, leaving it single stranded rather than double stranded. In the annealing step, the temperature is reduced, which allows the short oligonucleotide primers to find their complementary sequence on the template. The annealing temperature varies with primer length, primer sequence, and buffer condition of the reaction. In the extension stage, the heat-stable RNA polymerase adds nucleotides to the primers that have annealed to the template, generating two copies of the intervening DNA. By repeating the denature–anneal–extend cycle multiple times, exponentially more copies of the original target sequence that lies between the primers can be generated.

Recombination

http://www.genome.gov/glossary.cfm?key=homologous%20recombination

Recombination is the physical exchange of genetic material within a cell. This process represents a critically important mechanism for the repair of damaged DNA and for the generation of genetic diversity. Recombination occurs between homologous or repeated sequences after a nuclease or a DNA damaging agent nicks the phosphate backbone of one of strands. The nicked strand can then "probe" the sequences of other regions of the genome and can be extended by repair and replication proteins when a stable match for the nicked sequence is found. Since they are held in close proximity to one another, direct repeats of sequences are prone to recombination and consequent loss of the intervening DNA.

Restriction Enzyme

http://openwetware.org/wiki/20.109(F07):Start-up_genome_engineering

Restriction endonucleases, also called restriction enzymes, cut ("digest") DNA at specific sequences of bases. The restriction enzymes are named for the prokaryotic organism from which they were isolated. For example, the restriction endonuclease *Eco*RI (pronounced "echo-are-one") was originally isolated from *Escherichia* giving it the "Eco" part of the name. "RI" indicates the particular version on the *E. coli* strain (RY13) and the fact that it was the first restriction enzyme isolated from this strain.

The sequence of DNA that is bound and cleaved by an endonuclease is called the recognition sequence or restriction site. These sequences are usually four or six base pairs long and palindromic, that is, they read the same 5′ to 3′ on the top and bottom strand of DNA. Other restriction enzymes, for example *Hae*III, cut in the middle of the palindrome leaving no DNA overhang, called a "blunt end."

Phage Display

http://www.dyax.com/discovery/phagedisplay.html

The protein coat of bacteriophage can be exploited as a tool for discovery of protein–protein and protein–DNA interactions. With this technique, a protein encoded by the phage genome is modified with a randomized library of sequence. The modified genomes must express the modified coat protein so that the randomized portion is exposed on the phage surface. Genomes from the library generate a pool of phage each with a single peptide sequence exposed on the phage coat. The pool of phage can be selected for the binding properties of interest. Several "rounds" of selection, purification, and reselection are performed to isolate those phages with the desired binding properties. Phage candidates are finally analyzed through sequencing to identify the motif(s) from the library that are capable of binding the protein or DNA sequence of interest.

Author Biographies

Natalie Kuldell teaches in the Department of Biological Engineering at the Massachusetts Institute of Technology. She develops discovery-based curricula drawn from the current literature to engage undergraduate students in structured, reasonably authentic laboratory and project-based experiences. She completed her doctoral and postdoctoral work at Harvard Medical School, including experiments aimed at better understanding the regulated gene expression in the bacteriophage lambda. In 2005, she coedited a book on zinc-finger proteins with Shiro Iuchi, *Zinc Finger Proteins: From Atomic Contact to Cellular Function*. Her current research examines gene expression in the yeast, *Saccharomyces cerevisiae*, focusing on synthetic biology and improved tools for regulation of gene expression of the yeast mitochondria. Before joining the faculty of MIT, she taught at Wellesley College. Dr. Kuldell is the director of a web-based resource called BioBuilder to teach synthetic biology through comics and animations, as well as a scientific adviser for two web-projects to teach the nature and process of science, namely, Understanding Science and VisionLearning. She serves as Associate Education Director for SynBERC, an NSF-funded research center for Synthetic Biology, and Councilor at Large for the Institute of Biological Engineering. She is also one of three Boston-area coordinators for a public outreach network, the Coalition for the Public Understanding of Science.

Neal Lerner is Director of Training in Communication Instruction for the Program in Writing and Humanistic Studies at the Massachusetts Institute of Technology, where he teaches scientific and technical writing and supports lecturers and graduate students who teach in communications-intensive classes. Previous to MIT, he was a faculty member and Writing Programs Coordinator at the Massachusetts College of Pharmacy & Health Sciences. He earned an Ed.D. in Literacy, Language, and Cultural Studies from Boston University's School of Education and an MA in creative writing and a high school English teaching credential from San Jose State University. His current research is a history and contemporary classroom study of teaching high-school English in Holyoke, MA. He is a four-time recipient of the International Writing Centers Association Outstanding Scholarship Award and received the National Council of Teachers of English Award for Best Article Reporting Historical Research in Technical or Scientific Communication (2008).

He is the author of *The Idea of a Writing Laboratory* (Southern Illinois University Press, 2009), coauthor with Mya Poe and Jennifer Craig of *Learning to Communicate in Science and Engineering: Case Studies from MIT* (MIT Press, 2010), and coauthor with Paula Gillespie of *The Longman Guide to Peer Tutoring*, 2nd ed. (Longman, 2002). His publications have appeared in *College English, College Composition and Communication, Written Communication, IEEE Transactions on Professional Communication, Journal of Technical Writing and Communication, Writing Center Journal,* and several edited collections. He has held regional and national office in professional organizations for the study and teaching of writing and is a frequent consultant to colleges and universities in the United States and abroad on issues of teaching writing, writing program assessment, and writing across the curriculum.

Printed in the United States
by Baker & Taylor Publisher Services